Y0-EPI-404

# Selected Titles in This Series

669 **Seán Keel and James M^cKernan,** Rational curves on quasi-projective surfaces, 1999

668 **E. N. Dancer and P. Poláčik,** Realization of vector fields and dynamics of spatially homogeneous parabolic equations, 1999

667 **Ethan Akin,** Simplicial dynamical systems, 1999

666 **Mark Hovey and Neil P. Strickland,** Morava $K$-theories and localisation, 1999

665 **George Lawrence Ashline,** The defect relation of meromorphic maps on parabolic manifolds, 1999

664 **Xia Chen,** Limit theorems for functionals of ergodic Markov chains with general state space, 1999

663 **Ola Bratteli and Palle E. T. Jorgensen,** Iterated function systems and permutation representation of the Cuntz algebra, 1999

662 **B. H. Bowditch,** Treelike structures arising from continua and convergence groups, 1999

661 **J. P. C. Greenlees,** Rational $S^1$-equivariant stable homotopy theory, 1999

660 **Dale E. Alspach,** Tensor products and independent sums of $\mathcal{L}_p$-spaces, $1 < p < \infty$, 1999

659 **R. D. Nussbaum and S. M. Verduyn Lunel,** Generalizations of the Perron-Frobenius theorem for nonlinear maps, 1999

658 **Hasna Riahi,** Study of the critical points at infinity arising from the failure of the Palais-Smale condition for n-body type problems, 1999

657 **Richard F. Bass and Krzysztof Burdzy,** Cutting Brownian paths, 1999

656 **W. G. Bade, H. G. Dales, and Z. A. Lykova,** Algebraic and strong splittings of extensions of Banach algebras, 1999

655 **Yuval Z. Flicker,** Matching of orbital integrals on $GL(4)$ and $GSp(2)$, 1999

654 **Wancheng Sheng and Tong Zhang,** The Riemann problem for the transportation equations in gas dynamics, 1999

653 **L. C. Evans and W. Gangbo,** Differential equations methods for the Monge-Kantorovich mass transfer problem, 1999

652 **Arne Meurman and Mirko Primc,** Annihilating fields of standard modules of $\mathfrak{sl}(2,\mathbb{C})^\sim$ and combinatorial identities, 1999

651 **Lindsay N. Childs, Cornelius Greither, David J. Moss, Jim Sauerberg, and Karl Zimmermann,** Hopf algebras, polynomial formal groups, and Raynaud orders, 1998

650 **Ian M. Musson and Michel Van den Bergh,** Invariants under Tori of rings of differential operators and related topics, 1998

649 **Bernd Stellmacher and Franz Georg Timmesfeld,** Rank 3 amalgams, 1998

648 **Raúl E. Curto and Lawrence A. Fialkow,** Flat extensions of positive moment matrices: Recursively generated relations, 1998

647 **Wenxian Shen and Yingfei Yi,** Almost automorphic and almost periodic dynamics in skew-product semiflows, 1998

646 **Russell Johnson and Mahesh Nerurkar,** Controllability, stabilization, and the regulator problem for random differential systems, 1998

645 **Peter W. Bates, Kening Lu, and Chongchun Zeng,** Existence and persistence of invariant manifolds for semiflows in Banach space, 1998

644 **Michael David Weiner,** Bosonic construction of vertex operator para-algebras from symplectic affine Kac-Moody algebras, 1998

643 **Józef Dodziuk and Jay Jorgenson,** Spectral asymptotics on degenerating hyperbolic 3-manifolds, 1998

642 **Chu Wenchang,** Basic almost-poised hypergeometric series, 1998

641 **W. Bulla, F. Gesztesy, H. Holden, and G. Teschl,** Algebro-geometric quasi-periodic finite-gap solutions of the Toda and Kac-van Moerbeke hierarchies, 1998

*(Continued in the back of this publication)*

# Rational Curves on Quasi-Projective Surfaces

MEMOIRS
of the
American Mathematical Society

Number 669

# Rational Curves on Quasi-Projective Surfaces

Seán Keel
James McKernan

July 1999 • Volume 140 • Number 669 (third of 4 numbers) • ISSN 0065-9266

**American Mathematical Society**
Providence, Rhode Island

1991 *Mathematics Subject Classification*.
Primary 14J26, 14J45.

**Library of Congress Cataloging-in-Publication Data**

Keel, Seán.
Rational curves on quasi-projective surfaces / Seán Keel, James McKernan.
p. cm. — (Memoirs of the American Mathematical Society, ISSN 0065-9266 ; no. 669)
"July 1999, volume 140, number 669 (third of 4 numbers)."
Includes bibliographical references and index.
ISBN 0-8218-1096-0
1. Surfaces, Algebraic. 2. Algebraic varieties. I. McKernan, James, 1964– . II. Title. III. Series.
QA3.A57 no. 669
[QA573]
510 s—dc21
[516.3′52]

99-14985
CIP

## Memoirs of the American Mathematical Society

This journal is devoted entirely to research in pure and applied mathematics.

**Subscription information.** The 1999 subscription begins with volume 137 and consists of six mailings, each containing one or more numbers. Subscription prices for 1999 are $448 list, $358 institutional member. A late charge of 10% of the subscription price will be imposed on orders received from nonmembers after January 1 of the subscription year. Subscribers outside the United States and India must pay a postage surcharge of $30; subscribers in India must pay a postage surcharge of $43. Expedited delivery to destinations in North America $35; elsewhere $130. Each number may be ordered separately; *please specify number* when ordering an individual number. For prices and titles of recently released numbers, see the New Publications sections of the *Notices of the American Mathematical Society*.

**Back number information.** For back issues see the *AMS Catalog of Publications*.

Subscriptions and orders should be addressed to the American Mathematical Society, P. O. Box 5904, Boston, MA 02206-5904. *All orders must be accompanied by payment.* Other correspondence should be addressed to Box 6248, Providence, RI 02940-6248.



*Memoirs of the American Mathematical Society* is published bimonthly (each volume consisting usually of more than one number) by the American Mathematical Society at 201 Charles Street, Providence, RI 02904-2294. Periodicals postage paid at Providence, RI. Postmaster: Send address changes to Memoirs, American Mathematical Society, P. O. Box 6248, Providence, RI 02940-6248.


This publication is indexed in *Science Citation Index*®, *SciSearch*®, *Research Alert*®, *CompuMath Citation Index*®, *Current Contents*®/*Physical, Chemical & Earth Sciences*.
Printed in the United States of America.

∞ The paper used in this book is acid-free and falls within the guidelines established to ensure permanence and durability.
Visit the AMS home page at URL: `http://www.ams.org/`

10 9 8 7 6 5 4 3 2 1 04 03 02 01 00 99

## Contents

ABSTRACT

We verify a conjecture of Miyanishi about smooth quasi-projective surfaces over the complex numbers, namely, either such a surface has a pluricanonical log form, or a general point of the surface lies on a one dimensional image of the affine line. Applying standard techniques of the minimal model program, we reduce the problem to a detailed study of log del Pezzo surfaces. Log del Pezzo surfaces are normal projective surfaces, with ample anticanonical divisor and quotient singularities. We also give a partial classification of log del Pezzo surfaces of Picard number one.



1991 *Mathematics Subject Classification.* 14J26 and 14J45.

*Key words and phrases.* Log del Pezzo surface, log uniruled, quasi-projective surfaces, higher dimensional geometry, minimal model program, rational curves.

## §1 Introduction and Statement of Results

In classifying smooth projective varieties, one looks for an intrinsic map to projective space. The natural intrinsic line bundles are the various tensor powers of $\omega_X$, and so one is led to consider the pluricanonical linear series $|mK_X|$. The obvious question arises, what if these series are all empty? The answer is striking, one of the principal accomplishments of Mori's program:

**1.0 Theorem (Miyaoka, Mori, Kawamata, Shokurov, and others).** *Let $X$ be a smooth projective variety of dimension at most three.*

*Then $|mK_X|$ is empty for all $m > 0$ iff $X$ is covered by images of $\mathbb{P}^1$.*

(We say $X$ is covered by images of $\mathbb{P}^1$ if for every point $p$ of $X$, we may find a morphism $f : \mathbb{P}^1 \longrightarrow X$ such that the point $p$ belongs to $f(\mathbb{P}^1)$, where here, as in every other definition involving a morphism from a curve to a surface, we require $f$ to be non-constant.)

In this paper we consider the analogous question for quasi-projective (or more generally log) varieties. If $U$ is a smooth quasi-projective variety, then, following Iitaka, one picks a smooth compactification $U \subset X$ such that the complement $D = X \setminus U$ is a divisor with normal crossings. The linear series $|m(K_X + D)|$ turn out to depend only on $U$, not on $X$ or $D$, and is thus the natural analogue of $|mK_X|$. $|m(K_X + D)|$ is called the log pluricanonical series, and the problem is to characterise those smooth quasi-projective varieties for which the log pluricanonical series are all empty.

When $U$ is a curve the solution is elementary, $|m(K_X + D)|$ is empty for all $m > 0$ iff $U = \mathbb{A}^1$ or $\mathbb{P}^1$.

In dimension two the problem is already surprisingly subtle, and has received considerable attention. An import special case was settled by Miyanishi and Tsunoda, [31],[32], and further results have been obtained by Zhang, [36]. Here our main goal is a complete, and self-contained solution (throughout the paper, everything takes place over $\mathbb{C}$):

**1.1 Theorem.** *Let $U$ be a smooth quasi-projective variety of dimension at most two. Then $|m(K_X + D)|$ is empty for all $m > 0$ iff $U$ is dominated by images of $\mathbb{A}^1$.*

We say $U$ is **dominated by images of a curve** $C$, if there is a dense open subset $V$ of $U$ and for every point $p$ of $V$, we can find a non-constant morphism $f : C \longrightarrow U$ such that $p \in f(C)$. We say $U$ is dominated by rational curves, if it is dominated by images of $\mathbb{P}^1$.

Received by the editor September 20, 1997

We note that the reverse direction of the implication in (1.1) is fairly straightforward, and holds in all dimensions, see (5.11). Thus (1.1) can be viewed as saying, either $U$ has a log pluricanonical section, or there is a clear geometric reason why it cannot.

The other main results of this paper are very strong partial classifications of log del Pezzo surfaces, that is projective surfaces with quotient singularities and $-K_S$ ample. This includes a classification of all but a bounded family of log del Pezzo surfaces of Picard number one. We will explain this classification at the end of this introduction. We believe that with sufficient effort the methods of the paper would yield a complete classification.

Log del Pezzo surfaces are of interest for several reasons, independently of (1.1). They are obviously important for the study of open surfaces. Beyond this, the log category is important even for the study of projective varieties with, for example, log surfaces (surfaces with boundary) playing an intermediary rôle between surfaces and threefolds. Good examples of this are Kawamata's proof of the Abundance conjecture, and Shokurov's program for Flips. Log del Pezzos also occur as the centres of 4-fold log flips.

In addition log del Pezzos play an essential rôle in the compactification of the moduli space of surfaces of general type. They occur at the boundary of the moduli space of surfaces of general type (just as rational curves occur at the boundary of the moduli space of curves), and are the main object of study in the proofs of Alexeev's boundedness theorems [2], used to show the moduli space is projective.

We will use the following terminology. Given a variety $X$, let $X^0$ denote the smooth locus of $X$.

**1.2 Definition.** Let $X$ be a variety, $C$ a curve in $X$ and $D$ a subset of $X$. We say that $C$ meets $D$ $k$ times if the inverse image of $D$ on the normalisation of $C$ is a set of $k$ points.

For example a smooth point or a unibranch singularity of $C$ counts once, and more generally a singularity with $r$ branches counts $r$ times, regardless of the order of contact.

We note also that throughout the paper, by a rational curve, we mean a complete rational curve, that is an image of $\mathbb{P}^1$. By uniruled, we mean dominated by rational curves.

*1.2.1 Remarks.* If a complete variety is dominated by rational curves, then it is in fact covered by rational curves (that is through every point there is a rational curve), since the degeneration of a rational curve is again rational.

In the definition of dominating, one can equivalently require that the maps $f : C \longrightarrow U$ form a flat family (see (IV.1.3.5) of [26]).

**Connection between (1.1) and log del Pezzo surfaces.** In proving (1.1) one tries to simplify the situation by a birational contraction $\pi : X \longrightarrow Y$ of an irreducible curve $E \subset X$. In order to preserve the dimension of $|m(K_X + D)|$, it is sufficient that $(K_X + D) \cdot E < 0$. Under the contraction, the Picard number goes down by one, and from this point of view, $Y$ is simpler than $X$. The complication is that $Y$ can have quotient singularities (if $E$ has self-intersection at most $-2$). We replace $(X, D)$ by $(Y, D_Y = \pi_*(D))$, and continue, by looking for a $(K_Y + D_Y)$-negative contraction. Such a series of contactions is called the $(K_X + D)$-minimal model program (MMP for short).

If at the start $|m(K_X + D)| = \emptyset$ for all $m > 0$, then the Log Abundance Theorem implies that either $X$ has a fibration $X \longrightarrow C$ whose general fibre is $\mathbb{P}^1$ and meets $D$ at most once, or, we have a birational morphism $\pi : X \longrightarrow S$, a composition of contractions as above, with $S$ of Picard number one, and $K_S + D_S$ anti-ample and log terminal (discussed below). In the first case of course we take the general fibre. In the second case, $X$ is dominated by rational curves meeting $D$ at most once, iff $S$ is dominated by rational curves meeting $\pi(D)$ at most once. The one dimensional part of $\pi(D)$ is $D_S$, the zero dimensional part, $V$, is a union of components of $D$ contracted by $\pi$. In this way, (1.1) is reduced (and in fact equivalent) to:

**1.3 Theorem.** *Let $S$ be a normal projective surface of Picard number one, with quotient singularities. Suppose $D \subset S$ is a reduced curve, such that $K_S + D$ is log terminal, and $-(K_S + D)$ is ample. Let $V \subset S$ be any finite set of points. Then $S$ is dominated by rational curves, meeting $D \cup V$ at most once.*

An elementary discussion of the notion of log terminal is given in Appendix L. We note here (so that the reader may have some idea of the term's meaning) that if $D$ is irreducible then $K_S + D$ is log terminal iff the pair $(S, D)$ has quotient singularities, that is locally analytically the quotient of a smooth pair $(S', D')$ by a finite group. For the $(K_S + D)$-MMP, log terminal is the correct generalisation of normal crossings, in the sense that it is preserved by the birational contractions that occur in the $(K_S + D)$-minimal model program.

In fact in (1.3) it is enough to consider the case $V = \text{Sing}(S)$. The main point behind this reduction is Mori's observation that on a smooth space, the general member of a dominating family of images of $\mathbb{A}^1$ deforms freely (as an image of $\mathbb{A}^1$) and so in particular misses any fixed codimension two subset, see (5.5) and (5.8). Thus (1.3) can be equivalently (and somewhat more aesthetically) stated as:

**1.3.1 Theorem.** *Let $S$ be a normal projective surface of Picard number one, with quotient singularities. Suppose $D$ is a reduced curve, such that $K_S + D$ is log terminal, and $-(K_S + D)$ is ample. Then $S^0$ is dominated by rational curves which meet $D$ at most once.*

As we will explain shortly, the main work is proving (1.3) in the case when $D$ is empty, and is thus to show that the smooth locus of a log del Pezzo surface is uniruled (that is dominated by complete rational curves).

Our proof of (1.3) proceeds roughly as follows: For one class of rank one log del Pezzo surfaces, those with a tiger (defined below), we give a short and elegant deformation theoretic proof of (1.3), see (6.1). Boundedness results of Alexeev and Kollár imply that rank one log del Pezzos without a tiger are bounded, see (23.1). We complete the proof by constructing an explicit finite list of families of rank one log del Pezzos which includes any with no tiger and whose smooth locus has trivial algebraic fundamental group (it is easy to reduce (1.1) to this case, see §7). For each of the surfaces in this list, we directly construct a dominating family of rational curves.

We will explain all this in much greater detail below, but first we present some additional results which are of independent interest. Most are corollaries of (1.1). Proofs are given in §20. We will use the following notation.

$\mathbb{A}^1_* = \mathbb{A}^1 \setminus \{0\}$. In view of (5.11), it is natural to think of $\mathbb{A}^1$ and $\mathbb{A}^1_*$ as the open analogues of $\mathbb{P}^1$ and elliptic curves, since the existence of dominating rational, or elliptic families has analogous implications on ordinary Kodaira dimension.

**1.4 Proposition.** *Let $B \subset S$ be a reduced curve on a normal projective surface and set $U = S \setminus (B \cup \operatorname{Sing}(S))$. Consider the following conditions:*

(1) *The Kodaira dimension of $K_S + B$ is negative.*
(2) *$K_S + B$ is numerically trivial, but not log canonical.*
(3) *$K_S + B$ is numerically trivial, and $B \neq \emptyset$.*
(4) *$K_S$ is numerically trivial, $B = \emptyset$, and $S$ has a singularity which is* **not** *a quotient singularity.*

*If any of the above hold, then $U$ is dominated by images of $\mathbb{A}^1_*$, and by images of $\mathbb{A}^1$ if (1) or (2) holds.*

We note one interesting implication of (1.4.2) and (5.11): The complement of a integral plane cubic, $B$, is dominated by images of $\mathbb{A}^1$ iff $B$ has a cusp, or $B$ is the union of a smooth conic and a tangent line, or $B$ is the union of three lines meeting at a point (indeed this is just a list of the non log canonical possibilities).

Next we have a version of (1.3) for any boundary:

**1.5 Corollary.** *Suppose the pair $(S, \Delta)$ consists of a projective surface $S$ and boundary $\Delta$, such that $K_S + \Delta$ is log canonical. Then either*

(1) *$|m(K_S + \Delta)|$ is non-empty for some $m > 0$, or*
(2) *There is a covering family of rational curves $C$, such that $(K_S + \Delta) \cdot C < 0$.*

(1.5) thus says that either some multiple of $K_S + \Delta$ has a section, or there is a good geometric reason why it does not.

It may be tempting to believe that (1.5) follows automatically from the MMP: One can assume the $(K_S + \Delta)$-MMP gives a composition of contractions $f : S \longrightarrow S'$ such that $-(K'_S + \Delta')$ is ample. Every curve meets $(K_{S'} + \Delta')$-negatively. However, if the curve meets the exceptional locus of $f$, its strict transform may be $(K_S + \Delta)$-positive. Thus in order to prove (1.5) along these lines some result such as (1.3) is required. Such an attempt (in dimension three !) was the original impetus for this paper.

**1.6 Corollary.** *If $(S, \Delta)$ is a log Fano surface then $S^0$ is rationally connected, and $U = S \setminus \llcorner \Delta \lrcorner$ is connected by images of $\mathbb{A}^1_*$. In particular, $\pi_1(S^0)$ is finite, and $\pi_1(U)$ is almost Abelian.*

(Almost abelian is defined in §7.) The main issue in (1.6) is the uniruledness of the smooth locus of a rank one log del Pezzo. This was conjectured by Miyanishi and Tsunoda, [30]. The finiteness of the fundamental group $\pi_1(S^0)$ has been established previously in [14]-[15] and separately in [8]. The smooth locus of a log Fano threefold (or surface) has finite algebraic fundamental group by a boundedness result of [6].

**1.7 Corollary.** *Suppose $K_S + D + \Delta$ is log terminal for some effective $\mathbb{Q}$-divisor $\Delta$. Then every co-extremal ray (cf. [5]) is spanned by classes $[C]$, for $C$ the general member of a covering family of rational curves, contained in the smooth locus, and meeting $D$ in at most one point. In particular if $-(K_S + D + \Delta)$ is ample, then the nef cone of $S$ is polyhedral and spanned by such classes.*

The fact that the nef cone in (1.7) is polyhedral was proved in [5], our contribution is the description of the generators. (1.7) should be compared to the cone theorem, which describes generators for the cone of effective curves.

(1.7) in turn gives:

**1.8 Corollary.** *Suppose $(S, D + \Delta)$ is log Fano. Then $T_S(-\log D)$ is generically semi-positive, in other words for a general member $C$ of a sufficiently ample linear series, the vector bundle $T_S(-\log D)|_C$ has no quotient line bundles of negative degree.*

The fact that (1.8) is implied by (1.7) was pointed out to us by Mori.

We note that for terminal Fano threefolds, the analogue of (1.6) implies the analogue of (1.8). Thus if one could prove terminal Fano threefolds have uniruled smooth locus, it would follow as in [20] that the set of terminal Fano threefolds is bounded.

(1.8) itself has some interesting corollaries:

**1.8.1 Corollary.** *Let $(S, B)$ be a log Fano pair (with $S$ a projective surface, and $B \subset S$ a reduced curve).*

$$\sum_{p \in \mathrm{Sing}(S \setminus B)} \frac{r_p - 1}{r_p} \leq 2 + \rho(S) - e_{\mathrm{top}}(B)$$

*where $r_p$ is the order of the local fundamental group, $e_{\mathrm{top}}(B)$ is the topological Euler characteristic of $B$, and $\rho$ is the Picard number.*

Note when $B$ is empty then (1.8.1) implies that log del Pezzo surface can have at most $4+2\cdot\rho$ singularities. As far as we know, no such bound has been previously observed.

For $\rho = 1$, we prove a stronger version of (1.8.1) (it applies to any log terminal surface $S$), following an argument of [20], see (9.2). The proof is independent of (1.1) (and indeed, anything else in the paper). We call this intermediate result the **Bogomolov Bound** and use it repeatedly in the proof of (1.3). The Bogomolov Bound has an interesting corollary, see (9.3). In [2], using different methods, Alexeev obtains results related to (but much deeper than) (9.3).

In the light of (1.0) and (1.1), we propose the following:

**1.9 Conjecture.** *Let $U$ be a smooth quasi-projective variety. Then, either*

(1) $\bar{\kappa}(U) \geq 0$, *or*
(2) *$U$ is dominated by images of $\mathbb{A}^1$.*

Note that (1.9) is equivalent to the following:

**1.10 Conjecture.** *Let the pair $(X, D)$ consist of a smooth variety $X$, and reduced normal crossings divisor $D$. Then, either*

(1) $|m(K_X + D)|$ *is non-empty for some $m > 0$, or*
(2) *there is a dominating family of rational curves meeting $D$ in at most one point.*

By (5.11), (1) and (2) of (1.10) are mutually exclusive, and one can view the conjecture as saying either $X$ has a log pluricanonical section, or there is a clear geometric reason why none can exist.

There are natural analogues of (1.4-7) in higher dimensions, and similar implications will hold between them, if one has the MMP. We wish however to note one difference; in dimension two (1.1) implies (1.6). This follows from deformation theory, and the fact that log terminal surface singularities are quotient singularities, see Appendix L and §5. Three-fold log terminal singularities need not be quotient singularities.

In higher dimensions there is not much evidence for (1.9-10). We can prove (1.9) in the case $-(K_X + D)$ is ample, see (5.4), and a version of (1.3) for a threefold $X$ in the case $D$ has two components, see (6.7).

Under the assumptions of (1.9), if $X$ has dimension at most three, then the MMP and log Abundance [22] imply that $X$ is covered by (not necessarily rational) curves $C$ with $(K_X + D)\cdot C < 0$. Thus one is moved to consider:

**1.11 Conjecture: Log Bend and Break.** *(notation as in 1.10). If $(K_X + D)\cdot C < 0$, and $C \not\subset D$ then through a general point of $C$ there is a rational curve meeting $D$ at most once.*

Mori's famous bend and break argument proves (1.11) in the case $D$ empty. We don't see any way to extend the proof to the case when $D$ is not empty. In Mori's argument, $K_X \cdot C < 0$ is used to show that (mod $p$) some multiple of $C$ moves with a fixed point. If a curve moves with a point fixed, then rigidity implies there is a rational curve through the fixed point. However it is not clear (at least to us) how to control how the generated rational curve intersects $D$. In (6.9) we give examples indicating some of the difficulties.

In any case, the rational curves we use to prove (1.1) are obtained by an entirely different procedure, which we discuss below. However, once we have (1.1) we obtain a version of (1.11) as a:

**1.12 Corollary.** *Assumptions as in (1.10). If $X$ has dimension at most two, or $X$ has dimension three and the Kodaira dimension of $K_X + D$ is non-negative, then there is a rational curve through a general point of $C$, meeting $D$ at most once.*

**Outline of the proof of (1.3).**

Now we turn to an outline of the proof of (1.3), both to indicate its logical structure, and to high light many of issues which arise that are of independent interest. The logical order of the proof is also indicated by the flowchart (1.18) below.

The proof divides into two cases. We will use the notation of (1.3). Thus $S$ is a normal projective surface with Picard number one with quotient singularities. $D$ is a curve, such that $K_S + D$ is log terminal and $-(K_S + D)$ is ample.

Case I: $D \neq \emptyset$. The case of non-empty $D$ is proved in (6.2), using deformation theory and Kollár's Bug-Eyed cover. Here is a sketch: For simplicity, suppose $D$ is irreducible. By adjunction $D$ itself is a smooth rational curve, and we obtain a covering family of rational curves by deforming a high multiple of $D$. The difficulty of course is the presence of singularities. For this we use the Bug-Eyed cover: Given a normal surface $S$ with quotient singularities, there exists a unique smooth (but non-separated) algebraic space $\flat : S^\flat \longrightarrow S$ such that $\flat$ is a universal homeomorphism, and an isomorphism over $S^0$. See §4. For (1.3) the important point is that in terms of $\mathrm{Hom}(\Sigma, S^\flat)$, for a proper curve $\Sigma$, $S^\flat$ behaves exactly like a smooth variety. The idea then is to lift the problem to $S^\flat$. The condition that $K_S+D$ is log terminal is equivalent to the condition that $\flat^{-1}(D) \subset S^\flat$ is smooth, and the condition that $K_S + D$ is anti-ample, is equivalent to the existence of a endomorphism $f : \mathbb{P}^1 \longrightarrow D = \mathbb{P}^1$ and a commutative diagram

$$\begin{array}{ccc} \mathbb{P}^1 & \xrightarrow{\ g\ } & S^\flat \\ {\scriptstyle f}\downarrow & & \downarrow{\scriptstyle \flat} \\ D & \longrightarrow & S, \end{array}$$

see (4.14). Now deformation theory implies we can deform $g$ away from the singularities of $S$, maintaining a single point of (necessarily high order) contact with $D$.

The case of $D$ non-empty has been previously proved by Miyanishi and Tsunoda [31] and [32]. Their proof is based on a classification of pairs $(S, D)$. Deformation theory together with the Bug-Eyed cover gives a simple, transparent proof, as well as some generalisations to higher dimensions.

Case two: When $D$ is empty, the idea is to replace $S$ by $S_1$, birational to $S$, which contains a non-empty $D$, in such a way that (1.3) for $(S_1, D)$ implies $S^0$ is uniruled. $(S_1, D)$ will come from extracting a divisor $E$ via a blow up $f : T \longrightarrow S$, and then blowing down in a different direction $\pi : T \longrightarrow S_1$, $D = \pi(E)$. To find $(S_1, D)$ essentially reduces to finding $E$. For this we introduce the following, the most important technical definition of the paper:

**1.13 Definition.** Let $X$ be proper, normal variety. Let $\Delta$ be an effective $\mathbb{Q}$-Weil divisor on $X$. A **special tiger** for $K_X + \Delta$ is an effective $\mathbb{Q}$-divisor $\alpha$ such that $K_X + \Delta + \alpha$ is numerically trivial, but not Kawamata log terminal (klt).

For some $m$, $m\alpha$ is a very singular element of $|-m(K_X + \Delta)|$ and so a sort of antithesis to Reid's general elephant. Hence the terminology. By a special tiger for $X$, or a special tiger (without further reference), we mean a special tiger for $K_X$.

Note that if we have a special tiger there is at least one divisor $E$ of discrepancy (with respect to $K_S + \alpha$) at most $-1$. We will call any such $E$ a **tiger**. In fact we are much more interested in $E$ (and the resulting new log Mori fibre space structure) than in $\alpha$. Sometimes we will be a little sloppy in our notation and use the word tiger to mean either $E$ or $\alpha$. This should not cause any harm, because normally we are only interested in when $X$ does or does not have a tiger, $E$, which is equivalent to the existence of a special tiger, $\alpha$.

We will define klt, and explain the motivation behind (1.13) in (1.15) below.

The next two results indicate the usefulness of tigers for (1.3):

**1.14 Proposition.** *If $S$ is a projective surface with quotient singularities and $S$ has a tiger, then $S^0$ is uniruled.*

**1.15 Proposition.** *The collection of rank one log del Pezzo surfaces which do not have a special tiger is bounded.*

Note (1.14) is implied by (6.1).

(1.15) follows from (9.3) and quite general boundedness principals. We prove a stronger statement in §23. Of course, together Case I and (1.14-15) imply (1.3) in all but a bounded collection of cases. We will turn to the question of classifying these cases (that is surfaces without tiger) in a moment. But first some general remarks on tigers.

For the rest of the introduction, unless otherwise noted, $S$ will indicate a log del Pezzo surface of rank one.

Shokurov has independently considered tigers, in relation to complements. He has observed that if $S$ has a tiger, then $K_S$ is 1, 2, 3, 4 or 6-complemented ($K_S$ is $n$-complemented if there is a member $M \in |-nK_S|$ such that $K + 1/nM$ is log canonical). We include Shokurov's proof of this in §22. Note Shokurov's result, and (1.15) imply there is a uniform $N > 0$ such that $|-NK_S|$ is non-empty for all $S$. This result is interesting in view of the fact that there is no bound on the index of $K_S$.

Complements in dimension $n$ give information on extremal neighbourhoods in dimension $n+2$ (see [27]). They play an important role in Shokurov's program for log flips. These observations were pointed out to us by Alessio Corti.

Tigers have the following relation to the notion of affine-ruled (as defined by Miyanishi). The proof is given in (21.4).

**Lemma.** *If $S$ is affine-ruled (that is contains a product neighbourhood $U \times \mathbb{A}^1$) then $S$ has a tiger.*

There are $S$ with simply connected smooth locus (cf. §21), but no tiger. These give counter-examples to Miyanishi's conjecture (see [15] ) that the smooth locus of any rank one log del Pezzo has a finite étale cover, that is affine-ruled.

It is a fairly simple matter to reduce the proof of (1.3) to the case when $\pi_1^{alg}(S^0)$ is trivial (cf. §7). We will assume this for the rest of the introduction. We will sometimes abuse notation and say that $S^0$ (or even $S$) is simply connected. Of course, a posteriori, by (1.6), the two are even equivalent.

As remarked above, not every $S$ has a tiger. In §15-19 we generate a finite set of families of surfaces $\mathfrak{F}$ which includes all (simply connected) $S$ without tigers. For each $S \in \mathfrak{F}$ we have an explicit description of the minimal desingularisation, $\tilde{S}$, in terms of blow ups of $\mathbb{P}^2$, and we check in each case that $S^0$ is uniruled, by explicitly exhibiting a dominating family of rational curves. We note here that it is quite possible that $\mathfrak{F}$ is too big, in the sense that some of the surfaces in $\mathfrak{F}$ actually have tigers, or non-simply connected smooth locus. Given an explicit surface it is usually easier for us to check that the smooth locus is uniruled, than to check either that the smooth locus is simply connected, or that the surface has no tiger, as we have rather robust tools for the first, and only rather ad-hoc techniques for the second and third. Of course for (1.3) this approach is sufficient.

It is perhaps surprising, that even with an explicit description of $\tilde{S}$, it can still be very difficult to show $S^0$ is dominated by rational curves. Typically the strict transform of these covering curves have rather high degree in $\mathbb{P}^2$, often over a hundred.

To prove uniruledness in a specific case, we try to find rational curves, $Z \subset S$, with an endomorphism lifting to $S^b$. In practical situations this is only possible when $Z$ meets the singular locus at most twice, see (4.9.4). Then we try to deform the lifted endomorphism. We have the following sufficient condition (6.5):

Suppose $Z$ meets the singular locus twice, and the local analytic index of its two branches (meeting the singular locus) are $u$ and $v$. Then there is a surjection $f : \mathbb{P}^1 \longrightarrow Z$ of degree

$lcm(u, v)$ lifting to $S^b$, and if

$$-K_S \cdot Z \geq 1/u + 1/v$$

then $f$ deforms to cover $S^0$. We usually use the criterion when $Z$ has two branches meeting at one point, and $u = v$. In this case $uZ$ is Cartier, and so the criterion only fails when $-K_S \cdot Z = 1/u$. Example (6.8) indicates that the criterion is essentially sharp.

In order to apply the criterion to a particular case, we have to find rational curves meeting the singular locus twice. The $Z$ we use occasionally come from interesting geometric configurations (see for example 17.5.1-3). For concrete example applications of the criterion, see (6.10) and (8.1).

There are around sixty surfaces in $\mathfrak{F}$. The generation of $\mathfrak{F}$ is one main focus of the paper, and accounts for most of its volume.

**The hunt:** Our construction of $\mathfrak{F}$ is based on a simple idea. To explain it (and the notion of tiger) we use the following:

**1.16 Definition.** Let $X$ be a normal surface, and $\Delta$ an effective $\mathbb{Q}$-Weil divisor (that is $\Delta = \sum a_i D_i$ with $a_i \in \mathbb{Q}$, $a_i > 0$). Let $\pi : Y \longrightarrow X$ be any birational morphism. Let $\tilde{\Delta} = \sum a_i \tilde{D}_i$ be the strict transform. There is a unique $\mathbb{Q}$-Weil divisor $F$ supported on the exceptional locus such that

$$K_Y + \tilde{\Delta} + F = \pi^*(K_X + \Delta).$$

$\Gamma = \tilde{\Delta} + F$ is called the **log pullback** of $\Delta$. The **coefficient** $e(E, K_X + \Delta)$ of any irreducible divisor $E$ on $Y$ is just the coefficient as it appears in $\Gamma$, it depends only on $(X, \Delta)$ and the discrete valuation associated to $E$. The **coefficient** $e(X, \Delta)$ of the pair $(X, \Delta)$ is the largest coefficient of any divisor (or discrete valuation). We will write $e(X)$ for the coefficient of $(X, \emptyset)$.

*Remark-Definition.* $e(E, K_X + \Delta)$ is just the negative of its discrepancy. In particular, $K_X + \Delta$ is log canonical iff the coefficient of $(X, \Delta)$ is at most one, and log terminal iff in addition the coefficient of every exceptional divisor is less than one. It is Kawamata log terminal iff its coefficient is less than one.

We note one trivial, but useful, fact: *coefficients are invariant under log pullback.*

Now let us explain the motivation behind definition (1.13), specifically the definition of a special tiger for $S$. In view of case I, when $D$ in (1.3) is empty, one naturally wonders if there is some curve $C$ on $S$ with $K_S + C$ anti-ample and log terminal. It turns out this fails for infinitely many families of $S$, so one looks for weaker conditions. Note that if $K_S + C$ is anti-ample, then we can add on some effective $\mathbb{Q}$-Weil divisor $\beta$ so that $K_S + C + \beta$ is trivial. A general philosophy of the log category is to treat exceptional divisors and divisors on $S$ uniformly. Applying the philosophy to $\alpha = C + \beta$, leads to the notion of a special tiger (for $K_S$) which can be defined as an effective $\alpha$ with $K_S + \alpha$ numerically trivial, with coefficient at least one. We note also the following equivalent formulation of the definition of tiger (which follows readily from the definitions):

**Lemma.** *Let $E$ be an exceptional divisor over $S$. Let $T \longrightarrow S$ be the extraction (of relative Picard number one) of $E$. $E$ is a tiger iff $-(K_T + E)$ has non-negative Kodaira dimension.*

Since a tiger is a divisor of coefficient at least one, in hunting for a tiger, it is natural to extract the exceptional divisor from the minimal desingularisation which has maximal coefficient. This is the idea behind the hunt. We will use the tiger/hunt metaphor in various notations throughout the paper, occasionally to a tiresome degree.

Beginning with $(S_0, \Delta_0) = (S, \emptyset)$, we inductively construct a sequence of pairs $(S_i, \Delta_i)$ of a rank one log del Pezzo with a boundary, such that:

(1) $-(K_{S_i} + \Delta_i)$ is ample.

(2) If $(S_i, \Delta_i)$ has a tiger, then so does $S$.

The construction is by a sequence of a $K$-positive extraction (that is a blow up) $f_i : T_{i+1} \longrightarrow S_i$ followed by a $K$-negative contraction, $\pi_{i+1}$, each of relative Picard number one. $\pi_{i+1}$ is either a $\mathbb{P}^1$-fibration, or a blow down $\pi_{i+1} : T_{i+1} \longrightarrow S_{i+1}$. In the first case we say $T_{i+1}$ is a **net** and the process stops. We give the details in §8. Such sequences are frequently studied in the MMP, see for example [34]. The only choice in the sequence is which divisor is extracted by $f_i$. The hunt is a sequence given by extracting an exceptional divisor $E_{i+1}$, of the minimal desingularisation of $S_i$, for which the coefficient $e(E_{i+1}, K_{S_i} + \Delta_i)$ is maximal.

To generate the collection $\mathfrak{F}$ ( and complete the proof of (1.3) ) we classify all possibilities for the hunt for which we are unable to find a tiger. Let us give a few remarks to explain why such a classification is possible:

If the coefficient $e(S)$ is sufficiently close to one (cf. (21.1)), then by (5.4) of [25], $E_1$ is a tiger. By (9.3) the collection of $S$ with $e(S) < 1 - \epsilon$ is bounded. Thus in all but a bounded number of cases, the hunt finds a tiger at the very first step, and what is needed is an efficient means of dealing with the exceptions. Our choice for the hunt, that is always extracting a divisor of maximal coefficient (which is a natural choice, from the point of view of tigers) turns out to have remarkably strong geometric consequences. We will explain this in considerable detail in the introduction to §8. It is these consequences which make feasible an explicit classification of the exceptional cases. In (8.4.7) we give a detailed breakdown describing possibilities for the hunt. We then complete the proof by analysing each of the possibilities.

**1.17 Classification of all but a bounded collection of $S$.**

As discussed above, a detailed analysis of the hunt yields a collection $\mathfrak{F}$ containing all $S$ (with $S^0$ simply connected) without tiger. We introduced the hunt for exactly this purpose. Somewhat surprisingly, the hunt is also a useful tool for classification at the other extreme:

In §23 we classify rank one log del Pezzos $S = S_0$ (no assumption on fundamental group) such that $E_1$ (of the first hunt step) is a tiger. As we remarked above, this includes all $S$ with sufficiently large coefficient $e(S)$, which in turn includes all but a bounded collection of $S$.

Our classification, which is independent of the hunt analysis §14–19, is of the following sort:

First we classify abstract pairs $(S_1, A_1)$ of a rank one log del Pezzo surface containing an integral rational curve $A_1$ such that $K_{S_1} + A_1$ is anti-nef, and log terminal at singular points of $S_1$. They fall into a short number of series. We note that a similar classification of pairs is obtained in [31] and [32]. Our argument is based on quite different ideas, and is considerably simpler.

We apply this classification to the first hunt step. Let $A_1 \subset S_1$ be the image of $E_1$ (assuming $\pi_1$ is birational). It is easy to show that if $E_1$ is a tiger, then either $T_1$ is a $\mathbb{P}^1$-fibration, or $(S_1, A_1)$ is as in the preceding paragraph. The first case is easy to classify. To classify $S$, it remains to classify possibilities for the transformation $S_1 \dashrightarrow S$. This amounts to classifying possibilities for $\pi_1 : T_1 \longrightarrow S_1$ such that $E_1$ (the strict transform of $A_1$) is contractible, and contracts to a log terminal singularity. We indicate how this can be done in §23. It is easy and elementary, but we do not actually list the possibilities, as it would be notationally too involved.

Observe that the existence of such a classification is at least to some degree counter-intuitive. One might have expected a simple classification of $S$ with mild singularities, with a progressively less tractable list of possibilities as more complicated singularities are allowed. Indeed, $\mathbb{P}^2$ is the only smooth $S$, Gorenstein $S$ are classified in [11], and log del Pezzos with index (of $K$) at most two are classified in [1]. However, comparison of our two main classification results -the collection $\mathfrak{F}$, and our classification of $S$ with large coefficient-gives an indication in the opposite direction: Surfaces of small coefficient, though bounded, appear (at least to us) rather sporadic, while surfaces with sufficiently large coefficient exhibit relatively uniform behaviour.

Given the effectiveness of our techniques at either singularity extreme, we believe that repeated application of the same methods would eventually yield a complete classification of rank one log del Pezzo surfaces.

Several people have asked the following question; is it true that $\{K_S^2\}$ (over all log del Pezzos) satisfies ACC for bounded subsequences ? The question is suggested by analogy with a result of [2] that the corresponding set over minimal log terminal surfaces of general type satisfies DCC. The answer to this question is a resounding no, see (22.5).

Thanks: We would like to thank A. Baragar, A. Corti, D. Morrison, S. Mori, P. Deligne, D. Huybrechts, T. Hsu, M. Lustig, R. Morelli, Y. Petridis and V. Shokurov for advice and suggestions. We would like to particularly thank János Kollár, who suggested (1.5) as a natural companion to Log Abundance, and who has assisted us considerably throughout our research.

### 1.18 Logical Structure of Proof of (1.1)

Assume $\kappa(K_X + B) = -\infty$. Goal: $(X, B)$ log uniruled

Run $(K_X + B)$-negative MMP
$(X, B) \longrightarrow (S, D)$, $S$ rank one ldp

$\Downarrow$

(Goal: $(S, D)$ log uniruled; two cases:) $\Rightarrow$ If $D \neq \emptyset$: $(S, D)$ log uniruled (6.2) $\square$ $\Leftarrow$ {Bug-Eyed cover, §4, and Def. Theory §5}

$\Downarrow$

(If $D = \emptyset$: Goal $S^0$ uniruled; Two cases:) $\Rightarrow$ If $K_S$ has a tiger. $E$: transform $(S, \emptyset) \dashrightarrow (S_1, E)$ $S_1^0 \setminus E$ dominated by $\mathbb{A}^1_* s$ $\Rightarrow S^0$ uniruled. (6.1) $\square$ $\Leftarrow$ {Bug-Eyed Cover, Def. Theory, and Gorenstein ldps, §3}

$\Downarrow$

If $K_S$ has no tiger:
Run the Hunt

$\Downarrow$

$S \in \mathfrak{F}$
$\exists$ special rat $Z \subset S$ $\Rightarrow$ $S^0$ dominated by rats $\simeq mZ$ $\square$ $\Leftarrow$ {Def. Theory, Bug-Eyed cover, and criteria (6.5-6).}

$\Uparrow$

Analysis of Hunt, §8-19
See flow chart 8.0.16

## §2 Glossary of notation and conventions

If $S$ is a normal surface we indicate by $\tilde{S}$ its minimal desingularisation. If $C \subset S$ is an effective divisor, then $\tilde{C} \subset \tilde{S}$ will indicate its strict transform.

$\mathbb{F}_n = \mathbb{P}(\mathcal{O} \oplus \mathcal{O}(n))$ denotes the unique minimal rational ruled surface, with a curve $\sigma_\infty \subset \mathbb{F}_n$ of self-intersection $-n$, and $\overline{\mathbb{F}}_n$ denotes the log del Pezzo surface of rank one, obtained by contracting this curve. Let $\sigma_n \subset \mathbb{F}_n$ indicate a section disjoint from $\sigma_\infty$ (all such sections are linearly equivalent, and have self-intersection $n$).

Where no confusion arises, we will use the same notation to indicate a divisor, and its strict transform under a birational transformation. For a singular point $p \in S$, let $\rho(p)$ count the number of exceptional divisor of $\tilde{S} \longrightarrow S$ lying over $p$. On the other hand let $\rho$ be the relative Picard number of $\tilde{S} \longrightarrow S$. Clearly

$$\rho = \sum_{p \in S} \rho(p).$$

By a curve **over** a point $p \in S$, we mean a curve in the exceptional locus of $\tilde{S} \longrightarrow S$ over $p$.

By a $k$**-curve** on a normal surface $T$ we mean a complete curve whose strict transform on $\tilde{T}$ has self-intersection $k$. Note that if the curve is $K_T$-negative, and $k \leq 0$, then by adjunction, its strict transform is a smooth rational curve.

A $\mathbb{Q}$-Weil divisor $\sum a_i D_i$ is called a **subboundary** if $a_i \leq 1$, and a **boundary** if $0 \leq a_i \leq 1$. It is called **pure** if $a_i < 1$. $\sum a_i D_i$ is called reduced if all the non-zero $a_i$ are equal to one.

When we express a $\mathbb{Q}$-Weil divisor as $\sum a_i D_i$ we assume, that the $D_i$ are distinct and irreducible.

By a component of the $\mathbb{Q}$-Weil divisor $\Delta = \sum a_i D_i$, with $a_i \geq 0$, we mean a $D_i$ with $a_i > 0$. The support of $\Delta$, $\mathrm{Supp}(\Delta)$, is the union of its components (with reduced structure).

By $>$ or $<$ we mean strict inequality.

We write $\sum a_i D_i \geq \sum b_i D_i$ if $a_i \geq b_i$ for all $i$. We write $\sum a_i D_i > \sum b_i D_i$ if in addition $a_i > b_i$ for some $i$.

A log resolution of a pair $(X, D)$ of a variety $X$ and a reduced divisor $D$ is a birational map $f : Y \longrightarrow X$ with divisorial exceptional locus $E$, such that $Y$, and all components of $E + \tilde{D}$ are smooth, and $E + \tilde{D}$ has normal crossings. A log resolution of $(X, \Delta)$ is a log resolution of $(X, \mathrm{Supp}(\Delta))$. $(X, D)$ is called log smooth, if $X$ and every component of $D$ is smooth, and $(X, D)$ has normal crossings.

Throughout the paper, unless otherwise noted, by a divisor, we mean a $\mathbb{Q}$-Weil divisor, and by a component we mean an irreducible component.

We will make frequent, and occasionally unremarked use of the classification of two dimensional quotient singularities, see Chapter 3 of [27]. For the readers convenience we recall the most important facts in Appendix L. We also make frequent use of the formulas of Chapter 3 of [27] for the index and discrepancies. We will often refer to a cyclic singularity as a **chain** singularity, as the resolution graph is in this case a chain. For a non-cyclic singularity, there is exactly one vertex in the resolution graph meeting three edges. We call this the corresponding exceptional divisor the **central divisor**. Three chains meet this vertex. We refer to these as the **branches**. In a branch, we call the curve meeting the central divisor, the **adjacent** curve, and the other end of the chain, the **opposite** curve.

We will say that a chain singularity is **almost Du Val** if its resolution graph has only one vertex of weight other than 2, that of weight 3 and occurring at the end of the chain, that is the chain has form $(2, 2, 2, \ldots, 3)$ (where we also allow $(3)$).

By the index of a quotient singularity, we mean the order of the local fundamental group.

We will make frequent use of the main results of the (log) Minimal Model Program: the cone theorem, the contraction theorem, and the log abundance theorem. The first two of these are known in all dimensions, and the last up to dimension three. However we only use them in dimension two. For proofs in this case see [30]. For overviews of the general theory (as well as further references) see [16], [27], [21]. We will also make use of the standard definitions and notations of the program, as in [27].

We will say a pair $(X, D)$ of a normal variety and a reduced divisor is **log uniruled** if $X^0$ is covered by rational curves meeting $D$ at most once (that is through a general point of $x$ there is a complete rational curve, contained in the smooth locus, and meeting $D$ at most once). We will say $X$ is log uniruled if $(X, \emptyset)$ is log uniruled, or equivalently, if $X^0$ is uniruled, that is through a general point of $X$, there is a complete rational curve, contained in the smooth locus.

## §3 Gorenstein del Pezzo Surfaces

In this section we collect a number of results on Gorenstein log del Pezzo surfaces that we will use at various points in the paper. Especially useful will be (3.6-8) which together give a complete and simple picture of rank one Gorenstein log del Pezzo surfaces whose smooth locus is algebraically simply connected.

**Notation: Throughout the section $S$ denotes a rank one Gorenstein log del Pezzo.**

In [11] possible singularities of $S$ are classified. We will make frequent use of this classification, which we refer to as **the list**. We write for example $S(A_1 + A_3 + A_5)$ for a rank one Gorenstein del Pezzo with those singularities, and for example $\tilde{S}(A_1 + A_3 + A_5)$ for its desingularisation.

For the reader's convenience, here is a copy of the list:

(3.1)
$$\begin{array}{l} A_1, \quad A_1 + A_2, \quad A_4, \quad 2A_1 + A_3, \quad D_5, \quad A_1 + A_5, \quad 3A_2, \quad E_6, \\ 3A_1 + D_4, \quad A_7, \quad A_1 + D_6, \quad E_7, \quad A_1 + 2A_3, \quad A_2 + A_5, \\ D_8, \quad 2A_1 + D_6, \quad E_8, \quad A_1 + E_7, \quad A_1 + A_7, \quad 2A_4, \quad A_8, \\ A_1 + A_2 + A_5, \quad A_2 + E_6, \quad A_3 + D_5, \quad 4A_2, \quad 2A_1 + 2A_3, \quad 2D_4. \end{array}$$

By [33], the following are the subset of possibilities with $S^0$ algebraically simply connected, and any rank one Gorenstein del Pezzo with these singularities has algebraically simply connected smooth locus:

$$A_1, \quad A_1 + A_2, \quad A_4, \quad D_5, \quad E_6, \quad E_7, \quad E_8. \tag{3.2}$$

We will abuse notation and call this the **simply connected list.**

We note that in general $S$ is not determined by its singularities. However, for the simply connected list, this is almost the case. See (3.10) below.

The $K$-negative MMP for Gorenstein surfaces is almost as simple as that for smooth surfaces.

**3.3 Lemma.** *Let $g : T \longrightarrow W$ be a birational $K_T$-contraction, of relative Picard number one, with (irreducible) exceptional divisor $\Sigma$, such that $T$ is Gorenstein along $\Sigma$. Along $\Sigma$, $T$ is either smooth, or has a unique singularity. In the latter case the singularity is an $A_r$ singularity, with $K_T + \Sigma$ log terminal, and the singularity is removed by the contraction. Furthermore the induced map $\pi_1^{alg}(T^0) \longrightarrow \pi_1^{alg}(W^0)$ is an isomorphism.*

*Proof.* On the minimal desingularisation, $\Sigma$ is a $-1$-curve. Contractibility considerations give the description of the possible singularities. The last claim follows from (7.3). □

Gorenstein $\mathbb{P}^1$-fibrations (of relative Picard number one) are also quite simple. Here we state the possibilities. We will prove a more precise result in §11, see (11.5.4).

**3.4 Lemma.** *Suppose $\pi : T \longrightarrow C$ is a $\mathbb{P}^1$-fibration, of relative Picard number one, and $G \subset T$ is a fibre, contained in the Du Val locus of $T$. One of the following holds:*

(1) *$T$ is smooth along $G$.*
(2) *There are exactly two singularities, $A_1$ points, along $G$. $K_T + G$ is lt.*
(3) *There is a unique singularity, a $D_n$ point along $G$.*

*Proof.* Let $\tilde{T} \longrightarrow T$ be the minimal desingularisation, and let $h : \tilde{T} \longrightarrow W$ be a relative minimal model of $\tilde{T} \longrightarrow T$, thus $W \longrightarrow C$ is smooth, and $h$ is a composition of blow ups at smooth points. Because $\pi$ has relative Picard number one, each blow up in $h$ is at a point along the $-1$-curve of the previous blow up. Now one checks easily that the proposition gives all possibilities with Du Val singularities. For more details see (11.5). □

**3.5 Lemma.** *If $T$ is a Gorenstein del Pezzo surface, then $|\pi_1^{alg}(T^0)| \leq 8$.*

*Proof.* Note if $T' \longrightarrow T$ is a degree $n$ cover, étale in codimension one, then $T'$ is again a Gorenstein del Pezzo, and $K_{T'}^2 = nK_T^2 \leq 8$. Since $K_T^2$ is a positive integer, the result follows. □

Recall that throughout the paper, a proper $K_T$-negative curve on a normal surface $T$ is called a $-1$-curve, if $\tilde{C} \subset \tilde{T}$ is a $-1$-curve in the usual sense. Note when $T$ is Gorenstein, such a $C$ is a $-1$-curve iff $C$ is rational, $K_T \cdot C = -1$ and $\tilde{C}$ is smooth.

**3.6 Lemma.** *Let $S = S(E_8)$. $S$ contains a unique $-1$-curve $D$. $\tilde{D} \subset \tilde{S}$ meets a unique exceptional curve, the opposite end of the $A_4$ chain, with normal contact. $|-K_S|$ is one dimensional, and has a unique basepoint, a smooth point of $S$. There are two possibilities for the collection of rational members of $|-K_S|$. Either*

(1) *$|-K_S|$ has exactly three rational members, $D$ and two integral nodal rational curves $N_1$, $N_2 \subset S^0$, or*

(2) *$|-K_S|$ has exactly two rational members, $D$, and a unique integral cuspidal rational curve, $C \subset S^0$.*

*Proof.* Let $S = S(E_8)$ be any such surface. Obviously $S$ contains some $-1$-curve, $D$. $K_S^2 = 1$, so $|-K_S|$ is one dimensional by Riemann-Roch, and every member of $|-K_S|$ is reduced and irreducible. $|-K_S|$ and has a smooth elliptic member $E \subset S^0$ by [17]. $\mathcal{O}_E(E) = \mathcal{O}_E(q)$ for a unique $q \in E$. Since $H^1(\mathcal{O}_S) = 0$, $q \in S^0$ is the unique basepoint of $|-K_S|$. Let $T \longrightarrow S$ blow up $q$. $|-K_S|$ yields an elliptic fibration $g : T \longrightarrow \mathbb{P}^1$. Let $\tilde{g} : \tilde{T} \longrightarrow \mathbb{P}^1$ be the induced map. Since $S^0$ is simply connected, any $-1$-curve on $S$ is a member of $|-K_S|$. Since any $-1$-curve must pass through the singular point, and $g$ has irreducible fibres, $D$ is the unique $-1$-curve. The fibre of $\tilde{g}$ containing $D$ has 9 irreducible components, thus by Kodaira's classification of singular fibres, see page 150 of [4], the fibre containing $D$ is $\tilde{E}_8$, thus $D$ meets the opposite end of the $A_4$ chain as required. Note $e(\tilde{T})$ (the topological Euler characteristic) is 12. By formula (11.4) on page 97 of [4], the additional singular fibres (which we know are reduced and irreducible) are either exactly two nodal rational curves, or exactly one cuspidal rational curve. □

**3.7 Lemma.** *There are exactly two isomorphism classes of $S(E_8)$ corresponding to the two possibilities in (3.6).*

*Proof.* Let $S = S(E_8)$. Let $D$ be the unique $-1$-curve of (3.6). Let $B \subset S^0$ be any member of $|-K_S|$. Let $q \in B^0$ be the unique basepoint of $|-K_S|$, see (3.6). Let $L$ be the $-2$-curve of the $A_1$ branch of the $E_8$ singularity. Let $T \longrightarrow S$ extract $L$. $T$ has an $A_7$ singularity, and $D \subset T$ contracts $\pi : T \longrightarrow \mathbb{P}^2$, the image of $L$ is a flex line to the cubic curve $B \subset \mathbb{P}^2$ at (the image of the) $q$. The induced map $\tilde{S} \longrightarrow \mathbb{P}^2$ is obtained by blowing up 8 times along $B$ over $q$. $B \subset \mathbb{P}^2$ is embedded by the full linear system $|3q|$, thus $S$ depends only on $(B, q)$. The automorphism group of $B$ acts transitively on $B^0$, so $S$ depends only on $B$. By (3.6) we can take for $B$ either a cuspidal rational curve $C$, or a nodal rational curve $N$. □

**3.8 Lemma.** *Assume $S^0$ has trivial algebraic fundamental group. If $S$ is not $\mathbb{P}^2$ or $\overline{\mathbb{F}}_2$ then $S$ contains a unique $-1$-curve, $D$. $\tilde{D}$ has normal crossings with the exceptional locus of $\tilde{S} \longrightarrow S$, and meets exactly one exceptional curve (of the minimal desingularisation) over each singular point of $S$. If $S \neq \mathbb{P}^2$ then there is a $-2$-curve, $E$, of $\tilde{S}$, such that extracting $E$ gives a*

*$\mathbb{P}^1$-fibration, with $E$ a section, and $D$ (in case $S \neq \overline{\mathbb{F}}_2$) the only multiple fibre. $E$, and the singularities of $D$ are as follows:*

(1) *If $S = S(A_1) = \overline{\mathbb{F}}_2$ then $E$ is the unique $-2$-curve and the fibration is smooth.*
(2) *If $S = S(A_1 + A_2)$, then $K_S + D$ is log terminal, $E$ is the $-2$-curve over the $A_2$ point which is disjoint from $\tilde{D}$.*
(3) *If $S = S(A_4)$ with singularity $(2, 2', 2, \underline{2})$, then $D$ meets the primed curve and $E$ is the underlined curve.*
(4) *If $S = S(D_5)$ then $D$ meets one of the (2) branches, and $E$ is the opposite (from the central curve) end of the $A_2$ chain.*
(5) *If $S = S(E_k)$ $(8 \geq k \geq 6)$ then $D$ meets the opposite end of an $A_{k-4}$ chain (this chain is unique except when $k = 6$), and $E$ is the opposite end of an $A_2$ chain, different, in the case $k = 6$, from the $A_2$ chain which $D$ meets.*

*Furthermore, assume $1 < K_S^2 < 8$. Let $T \longrightarrow \tilde{S}$ blow up a point of $D$ not on any $-2$-curve. Then $T$ is the minimal desingularisation of a rank one log del Pezzo, $S'$ with algebraically simply connected smooth locus, and $K_{S'}^2 = K_S^2 - 1$. If we repeat this process $K_S^2 - 1$ times, we obtain $S(E_8)$. The induced map $\tilde{S}(E_8) \longrightarrow \tilde{S}$ is canonical, contracting at each stage the unique $-1$-curve.*

*Proof.* We will prove that there exists a $-1$-curve, $D$, meeting the singularities as prescribed. The final remarks are then immediate from the singularity description, and imply the uniqueness of the $-1$-curve by (3.6). One also checks easily that extracting the indicated $-2$-curve $E$ gives the required $\mathbb{P}^1$-fibration. Hence it is enough to prove the existence of $D$.

We can assume by (3.6) that $K_S^2 \geq 2$.

Note if $C \subset S$ is any $-1$-curve, then $(K_S + C) \cdot C < 0$ and thus $C$ is smooth, and so meets at most one exceptional divisor over each singular point, and the contact is normal (see for example (6.11)).

Let $f : T \longrightarrow S$ extract a $-2$-curve which according to the statement of the lemma is to have contact with $D$. In the case of $S(A_1 + A_2)$ let $V$ be the $-2$-curve of the $A_1$ point, otherwise choose any of the possible curves, that is on $S(D_5)$ either of the $A_1$ branches, on $S(E_6)$ the opposite end of either $A_2$ chain, and on $S(A_4)$ either of the interior $-2$-curves (in the other cases $V$ is unique). Suppose first that $T$ has a $\mathbb{P}^1$-fibration. By (3.4) this is only possible if $S = S(E_6)$. In this case one checks that the fibre through the $D_5$ point of $T$ is a $-1$-curve meeting the singularities as prescribed (for details see (11.5.4)). So we can assume $T$ has a birational contraction $\pi : T \longrightarrow S_1$ of a $-1$-curve, $D$. Using (3.3) and the list, one checks in each case that $D$ meets the singularities as prescribed:

If $S = S(D_5)$, $T$ has an $A_4$ singularity. By (3.3) and the simply connected list $S_1$ is either $\mathbb{P}^2$, $D \subset T$ contains the $A_4$ point, and $K_T + D$ is log terminal, or $S' = S(A_4)$ and $D \subset T^0$. In the second case $D \subset S$ has the prescribed singularities. In the first either the singularities are as required, or $D$ meets the opposite end of the $A_3$ chain. But the latter is impossible, for then $D^2 = 0$ on $S$.

If $S = S(A_4)$. $T$ has singularities $A_1 + A_2$. Either $D \subset T^0$, and $S_1 = S(A_1 + A_2)$, or $D$ meets one end of the $A_2$ chain, and $S_1 = S(A_1)$. In the first case, either the singularities are as prescribed, or $D \subset S$ meets an end of the $A_4$ chain. But the latter is impossible, for in that case $D^2 < 0$ on $S$.

If $S = S(A_4)$, $T$ has a single singularity, an $A_2$ point. $D \not\subset T^0$, from the list, so $D \subset S$ has the prescribed singularities.

In the cases, $S(E_6)$, $S(E_7)$, $T$ has a single singularity, a non-cyclic singularity, so by (3.3) $D \subset T^0$, and the singularities are as prescribed. □

**3.9 Corollary.** *Let $W$ be a Gorenstein log del Pezzo surface of rank one, which contains two distinct rational curves $D_1$ and $D_2$ with $K_W \cdot D_i = -1$. The following implications hold:*

(1) *If $W$ is simply connected then $W = S(E_8)$, one of the $D_i$ is contained in $W^0$, and is a rational curve of arithmetic genus one. $D_1 \cap D_2$ is a single smooth point of $W$, the unique basepoint of the linear series $|-K_W|$.*

(2) *If $K_W^2 \geq 4$, then $K_W^2 = 4$, $W = S(2A_1 + A_3)$ and $D_1$ and $D_2$ each pass through one of the $A_1$ points and opposite ends of the $A_3$ point.*

*Proof.* (1) is immediate from (3.6), (3.7) and (3.8).

Now suppose $K_W^2 \geq 4$. By (1) $W$ is not simply connected, and so from the (full) list, $S = S(2A_1 + A_3)$. $K_W + D_1 + D_2$ is anti-ample. In particular $D_i$ is smooth. By adjunction, they meet at only one point, and neither can contain all three singular points. Thus each $D_i$ must contain exactly two singular points, and they cannot both be $A_1$, for otherwise $D_i^2 \leq 0$. Thus each $D_i$ contains one of the $A_1$ points, and they both contain the $A_3$ point. The described configuration is now the only possibility, as one checks that otherwise after extracting an appropriate curve over the $A_3$ point, both have non-positive self-intersection, and at least one is contractible, violating the Picard number. □

**3.10 Corollary.** *Let $S$ be a rank one Gorenstein log del Pezzo surface, with algebraically simply connected smooth locus. $S$ is uniquely determined by its singularities (and thus by $K_S^2$) except in the case of $S(E_8)$ (that is $K_S^2 = 1$) when there are two possibilities, as described in (3.7).*

*Proof.* We will write $S(C)$ (resp $S(N)$) for the $S(E_8)$ of (3.6-7) with a cuspidal member (resp. nodal member).

$S(A_1)$ is unique by (3.8). Assume $1 < K_S^2 < 8$.

By the final remarks of (3.8) there are at most two possibilities for $\tilde{S}$: Start with either $S(C)$ or $S(N)$ and contract repeatedly the unique $-1$-curve, $K_S^2 - 1$ times. We will show that the two possibilities give the same surface. Start with $S(C)$. One obtains $S$ with $C \subset S^0$, $C \in |-K_S|$. Let $f : \mathbb{P}^1 \longrightarrow C$ be the normalisation. $\mathrm{Hom}(\mathbb{P}^1, S)$ has dimension at least 4 at $f$ by (II.1.2) of [26], and thus deforms to give a dominating family of rational curves. By (II.3.14) of [26] the general member has only nodes, and so gives a nodal rational curve $N \in |-K_S|$, $N \subset S^0$. Choosing the point $x$, of the final remarks in (3.8), to be (repeatedly) $N \cap D$, we see we can also obtain $S$ starting with $S(N)$. $\square$

## §4 Bug Eyed Covers.

**Brief Introduction.**

As indicated in the introduction, a key tool in the proof of (1.1) is Kollár's bug-eyed cover. For a rank one log del Pezzo $S$, this is a universal homeomorphism $\flat : S^\flat \longrightarrow S$, with $S^\flat$ a non-singular (but if $S$ is singular, non-separated) algebraic space. Informally, $S^\flat$ allows us to do deformation theory on $S$, *as if $S$ were smooth.* More precisely, we have a good deformation theory for a map $f : C \longrightarrow S$ (from a proper smooth curve) provided $f$ factors through $\flat$. Here our main goal is to present criteria for the existence of such a lifting. We will consider the question of how to deform $f$ in §5 and §6.

We note at the outset that all of the important ideas in this section are due to Kollár, and most of the statements below can be found implicitly or explicitly in [23]. We present them here rather than quote Kollár directly, as it will be convenient to have them in a slightly different form. Our only original contribution to the general theory is the uniqueness of the bug-eyed cover of a variety with isolated quotient singularities, (4.4). This is stated in [23], but without proof.

The reader who is familiar with Kollár's construction, and willing to take uniqueness on faith, can skip directly to (4.5).

One point of notation: or us a curve is assumed to be a separated variety. Recall that everything in this paper takes place over $\mathbb{C}$.

We begin by recalling Kollár's definition:

**4.0 Definition.** Let $f : Y \longrightarrow X$ be a map between algebraic spaces. Let $U \subset Y$ be the locus where $f$ is étale. Assume $U$ is separated. The **bug-eyed cover of $X$ associated to $f$** is the algebraic space $Y/R_f \longrightarrow X$, where $R_f$ is the equivalence relation $R_f = (U \underset{X}{\times} U \setminus \Delta_U) \cup \Delta_Y$. Let $B = U^c$.

*4.1 Remark.* Informally, $Y/R_f \longrightarrow X$ is obtained from $Y \longrightarrow X$ by gluing together inverse image points where $f$ is étale. In particular, if $f$ is étale at every inverse image of $x$, then $Y/R_f \longrightarrow X$ is an isomorphism over a neighbourhood of $x$. It is a bit surprising to us that such a simple idea turns out to be so useful.

**4.2 Lemma.** *Let $g : Y' \longrightarrow Y$ be a separated étale surjection, which is one to one on closed points over $B$. Then*

(1) $R_{f\circ g} = R_f \underset{Y\times Y}{\times} (Y' \times Y')$,

(2) *the induced map $Y'/R_{f\circ g} \longrightarrow Y/R$ is an isomorphism, and*

(3) *if further $f$ is an isomorphism on $U$ and $f$ is bijective on closed points, then $Y \longrightarrow X$ is the bug-eyed cover associated to $f \circ g$, $Y'/R_{f\circ g}$.*

*Proof.* Let $\amalg$ indicate a disjoint union into open and closed subsets. Note $U' = g^{-1}(U)$ ($U', B' \subset Y'$ are defined with respect to $f \circ g$ in the obvious way). Since $g$ is separated and étale, there is a decomposition $Y' \underset{Y}{\times} Y' = A \amalg \Delta_{Y'}$. By assumption $B' \underset{Y}{\times} B' \subset \Delta_{Y'}$, and so $U' \underset{Y}{\times} U' = A \amalg \Delta_{U'}$. Thus we have

$$\begin{aligned} R_f \underset{Y\times Y}{\times} (Y' \times Y') &= (U' \underset{X}{\times} U' \setminus U' \underset{Y}{\times} U') \amalg Y' \underset{Y}{\times} Y' \\ &= (U' \underset{X}{\times} U' \setminus \Delta_{U'}) \amalg \Delta_{Y'} = R_{g\circ f}. \end{aligned}$$

Hence (1). Since $g$ is an étale surjection, (2) follows.

Note that when $f$ is an isomorphism on $U$ and $f$ is bijective on closed points, $R_f = \Delta_Y$. This gives (3). $\square$

Now we turn to uniqueness. We first consider what happens locally. Note that when $X$ is two dimensional, the case that most interests us, the local analytic possibilities for $f$ are classified in [7].

**4.3 Lemma.** *Let $(X, x)$ be a pointed normal variety of dimension $n \geq 2$. Let $f : (W, w) \longrightarrow (X, x)$ be a quasi-finite surjection from a smooth algebraic space, with $w = f^{-1}(x)$ (as sets), étale away from $w$. Let $G$ be the local fundamental group of $X$ at $p$.*

(1) *If $W$ is a separated variety, then locally analytically around $x$ and $w$,*

$$f : W \setminus \{w\} \longrightarrow X \setminus \{x\}$$

*is the universal cover.*

(2) *The extension of Hensilisations $\mathcal{O}^h_{X,p} \subset \mathcal{O}^h_{W,w}$ induces a Galois extension of quotient fields, with group $G$, and $\mathcal{O}^h_{X,p} = (\mathcal{O}^h_{W,w})^G$.*

*Proof.* Replacing $W$ by an étale cover with a single point lying over $w$, (which doesn't effect $\mathcal{O}^h_{W,w}$) we may assume $W$ is an affine variety. Locally analytically $f : W \setminus \{w\} \longrightarrow X \setminus \{x\}$ is a cover, and since $W \setminus \{w\}$ is simply connected (locally analytically), it is the universal cover. Hence (1). The analogous statement to (2), for analytic local rings follows. (2) now follows by (4.11) of [3]. $\square$

**4.4 Proposition.** *Let $X$ be a normal variety with isolated quotient singularities. Then there is a unique smooth algebraic space $f : W \longrightarrow X$ such that $f$ is an isomorphism over $X^0$, and one to one on closed points. The map of Hensilisations is described by (4.3.2). $f$ is a universal homeomorphism. In particular, when $X$ is complete, $W$ is universally closed.*

*Proof.* By definition there is a surjection $Y \longrightarrow X$, étale outside the singular locus, where $Y$ is a smooth variety. By dropping points of $Y$ we can assume the map is one to one on points over the singular locus. For the existence of $W$ we take the associated bug-eyed cover.

Uniqueness can be checked Zariski locally on $X$. Thus for the rest of the proof we work in the category of pointed spaces. Suppose $f_i : (W_i, w_i) \longrightarrow (X, x)$, $i = 1, 2$, are two candidates. By (4.2) there is a smooth affine variety $(W, w)$ and a pair of étale surjections $p_i : (W, w) \longrightarrow (W_i, w_i)$. Dropping closed points, we can assume $w = p_i^{-1}(w_i)$. Since $X$ is separated, $f_1 \circ p_1 = f_2 \circ p_2$. By (4.2.3), $W_i = W/R_{f_i \circ p_i}$.

Since $f$ is a set bijection, to show it is a universal homeomorphism, it is enough to show it is universally open. This is a local étale question on $W$ and $X$, thus we reduce to the case of a quotient by a finite group, where the result is known. Alternatively, one can apply Chevalley's criterion, see (15.4) of [13]. □

**4.5 Definition.** We will denote the unique $W$ of (4.4) by $\flat : X^\flat \longrightarrow X$, and refer to it as **the bug-eyed cover of** $X$.

We will be principally interested in the bug-eyed cover of a log del Pezzo surface, $S$.

Next we turn to lifting criteria.

**4.6 Lemma.** *Let $c : C' \longrightarrow C$ be a surjective generic isomorphism from a non-singular one dimensional algebraic space to a smooth curve. For each $Q \in C'$ such that $c$ is not an isomorphism over $p_Q = c(Q)$, let $r_Q$ be the ramification index. Then in the notation of [23], $C' = C\{r_Q, p_Q\}$.*

*Proof.* Let $D = C\{r_Q, p_Q\}$. There is an obvious set bijection between $D$ and $C'$ and one can check locally that this is an isomorphism. Thus one can drop points from fibres, and assume $c$ is a set bijection. In this case the result follows from (4.2.3) above, and (4.3) of [23]. □

We now present Kollár's basic lifting criterion in a convenient intrinsic form:

**4.7 Lemma.** *Let $c : C' \longrightarrow C$ be as in (4.6). Let $f : D \longrightarrow C$ be a finite map, with $D$ a smooth curve. Then $f$ factors through $c$ iff the following holds: for each point $y \in D$, there is an inverse image point $x \in c^{-1}(f(y))$ such that the ramification index $r_c(x)$ divides the ramification index $r_f(y)$.*

*Proof.* Immediate from (4.6), and (4.2.2) of [23]. □

**4.8 Lemma.** *Let $C'$ be a smooth one dimensional (not necessarily separated) algebraic space. Let $C$ be a proper smooth curve, and let $c : C' \longrightarrow C$ be a generically isomorphic surjection.*

*If $K_{C'}$ has negative degree then $C = \mathbb{P}^1$ and there exists an endomorphism $f$ of $C$ which factors through $c$.*

*If further $c$ is a set bijection, then the converse holds. If $C = \mathbb{P}^1$ and there is an endomorphism $f$ factoring through $c$, then $K_{C'}$ has negative degree.*

*Proof.* By (4.6), $C' = C\{r_i, p_i\}$. $K_{C'}$ has degree $-2 + \sum \frac{r_i - 1}{r_i}$ by the Hurwitz formula. Now apply (4.4.2) and (6.5) of [23] (note the remark at the end of the proof of (6.5)). $\square$

*4.9 Remarks.*

(1) Notation as in (4.8). We note that by (4.6) above, (4.4.2) and (6.5) of [23] there is always a quasi-finite surjection $d : D \longrightarrow C'$ with $D$ a proper smooth curve, and the degree of a line bundle $L$ on $C'$ can be defined (as a rational number) as $\deg(d^*(L))/\deg(d)$.

(2) Of course in (4.7) the condition on $f(y)$ holds if there is a point over $f(y)$ where $c$ is étale. Thus one only needs to consider points in $C$ over which $c$ is *no where étale.*

(3) (4.7) implies in particular that if $C = \mathbb{P}^1$ and there are at most two points in $C$ over which $c$ is no where étale, then there is an endomorphism $f$ of $C$ factoring through $c$.

(4) In the final paragraph of (4.8), when $C = \mathbb{P}^1$, one can express the necessary and sufficient condition for the existence of a lifting endomorphism as: There are at most three points in $C'$ where the map is not étale and if $r_1$, $r_2$, $r_3$ are the ramification indices (which are allowed to be one), then there exists a lifting endomorphism iff $2 > \sum \frac{r_i - 1}{r_i}$. For three ramification points, the possible $(r_1, r_2, r_3)$ are the so called *Platonic triples*, ubiquitous in the study of lt surface singularities.

**4.10 Lemma.** *Let $f : C \longrightarrow S$ be a proper map, and a generic embedding, where $C$ is a smooth curve and $S$ is a rank one log del Pezzo surface. Let $C'$ be the normalisation of $\flat^{-1}(f(C))$ (for the construction of $C'$ see Appendix N).*

*(1) There is a commutative diagram*

$$\begin{array}{ccc} C' & \longrightarrow & S^\flat \\ {\scriptstyle c}\downarrow & & \downarrow \\ C & \xrightarrow{\ f\ } & S. \end{array}$$

*(2) If $D \longrightarrow C$ is a map from a smooth curve such that $D \longrightarrow S$ factors through $h : D \longrightarrow S^\flat$, then $h$ factors uniquely through $C'$.*

*Let $p \in S$ and $g : (U, y) \longrightarrow (V, p)$ express a local analytic neighbourhood $V$ of $p$ as a quotient of smooth $U$ by the finite group $G$ (so $G$ acts on $U$ freely outside of $y$, and $G$ is the local fundamental group). Fix $q \in f^{-1}(p)$. Let $Z$ be the corresponding local analytic branch of $f(C)$*

*through $p$. Let $n$ be the number of analytic branches of $g^{-1}(Z)$. Let $m$ be the index of $S$ at $p$ (that is the order of $G$) and let $d$ be the local analytic index of $Z$. Let $t \in C'$ be a point of $c^{-1}(q)$.*

*(3) $n$ is the number of points in $c^{-1}(q)$. $n|m$ and $r = m/n$ is the ramification index of $c$ at $t$. $d|r$. If $m$ is prime then $d = r = m$ unless $Z$ is Cartier at $p$.*

*Proof.* (1) and (2) are immediate from (N.3), the universal property of normalisation.

(3) can be considered at the level of strict Hensilisations and so by (4.3) we can replace $\flat : S^\flat \longrightarrow S$ by $g : (U, y) \longrightarrow (V, p)$ and $C'$ by the normalisation of $g^{-1}(Z)$.

By (5.E.vi) of [28] $G$ acts transitively on the branches of $g^{-1}(Z)$, and so $n|m$. Clearly each branch of $C'$ has degree $m/n$ over $C$, thus $r = m/n$, since $q$ has a unique inverse image on each branch.

Let $h$ be a local equation for one branch of $g^{-1}(Z)$. Since $G$ acts transitively on the branches of $D$, $N_G(h)$ (the product of the translates), is a defining equation for $r \cdot g^{-1}(Z)$. Since $N_G(h)$ is $G$ invariant, it's a defining equation for $rZ \subset V$. Thus $rZ$ is Cartier, and so $d|r$ by the definition of the index.

Finally, if $m$ is prime, and $d \neq m$, then $d = 1$, and $Z$ is Cartier. □

It will be convenient for applications to have a criterion for when $Z$ of (4.10) fails to be Cartier.

**4.11 Lemma.** *Let $(A, m)$ be a one dimensional local Noetherian Domain, a $k$-algebra, with $k$ isomorphic to $A/m$. Assume the normalisation $\tilde{A}$ is finite over $A$ and local. Then*

$$\dim_k \frac{m}{m^2} \leq \min_{h \in m \setminus 0} \dim_k \frac{A}{h} = \dim_k \frac{\tilde{A}}{m \cdot \tilde{A}}.$$

*Proof.* By the projection formula [9]

$$dim_k \frac{A}{h} = \dim_k \frac{\tilde{A}}{h \cdot \tilde{A}}.$$

Since $\tilde{A}$ is a local DVR, the second equality follows.

$$dim_k \frac{m}{m^2} \leq dim_k \frac{m}{m^2 + h} + 1 \leq dim_k \frac{m}{h} + 1 = \dim_k \frac{A}{h}$$

which gives the first inequality. □

**4.12 Lemma.** *Let $(S, p)$ be a germ of a normal surface, with a rational singularity, and $g : \tilde{S} \longrightarrow S$ the minimal desingularisation, with reduced exceptional divisor $E$. Let $p \in C \subset S$ be an analytically irreducible curve, such that the strict transform $\tilde{C} \subset \tilde{S}$ is smooth. Assume $E$ is $g$-anti-nef.*

(1) $\dim T_p(C) \leq \tilde{C} \cdot E$.

(2) *$C$ is smooth iff $\tilde{C} \cdot E = 1$. In particular if $\tilde{C} \cdot E = 1$ then $C$ is not Cartier.*

(3) *If $\tilde{C} \cdot E = 2$ then $C$ has a hypersurface singularity. If furthermore $C \subset S$ is Cartier, then $S$ is Du Val.*

*Proof.* Since $E$ is anti-nef, there is a sum of analytic discs $W$, meeting $E$ normally, and disjoint from $\tilde{C}$ such that $E + W$ is cut out by $g^{-1}(h)$ for some $h \in m_{S,p}$. (1) follows from (4.11).

Now suppose $C$ is smooth. Then if we take $h$ a generator of $m_{C,p}$, then

$$1 = Z(h) \cdot \tilde{C} \geq E \cdot \tilde{C}$$

(where $Z(h) \subset \tilde{S}$ is the scheme cut out by $g^*(h)$) which gives (2).

For (3) suppose $\tilde{C} \cdot E = 2$. Then by (1-2), $C$ has two dimensional tangent space and in particular $C$ is Gorenstein. Now if $C$ is Cartier, then $S$ is Gorenstein. Hence (3). □

*4.13 Remarks.* Note the conditions of (4.12) hold for any cyclic quotient singularity, and for any non-cyclic quotient singularity for which the central curve has self-intersection at most $-3$. The implication $C$ smooth $\implies$ $\tilde{C} \cdot E = 1$ (and its proof) holds for any curve on a normal surface germ, that is without assuming $E$ is $g$-anti-nef. The reverse implication, however, does not. One counter example is the unique $-1$-curve on $S(E_8)$, see (3.8).

*4.14 Remark.* The conditions for a pair $(S, D)$ of an irreducible (non-empty) curve on a rank one log terminal surface to be log Fano have a nice expression in terms of the bug-eyed cover $\flat : S^{\flat} \longrightarrow S$:

(1) An irreducible curve $D$ on a log terminal surface $S$ is log terminal iff the pullback $\flat^*(D) \subset S$ is non-singular.

(2) If $D$ is log terminal and $S$ has rank one, then $K_S + D$ is anti-ample iff $D = \mathbb{P}^1$ and there exists an endomorphism of $D$ which factors through $\flat$.

*Proof.* For (1) see Appendix L. (2) follows from (4.8) (or (4.9.3)) and the adjunction formula, see (4.15) below. □

*4.15 Remark.* For log terminal singularities, Shokurov's different (see Appendix L) also has a nice interpretation in terms of the bug-eyed cover. Suppose we have an lt pair $(S, C)$, with $C$ reduced and irreducible. Suppose the singular points of $S$ along $C$ are $p_1, p_2, \dots, p_n$ and $r_1, r_2, \dots, r_n$ are the corresponding orders of the local fundamental groups. $C' = \flat^{-1}(C)$ is non-singular as $C$ is lt, and $c : C' \longrightarrow C$ is a set bijection, totally ramified at each $p_i$ with ramification index $r_i$. In the notation of [23], $C' = C\{r_i, p_i\}$. Let $q_i$ be the unique point of $D$

over $p_i$. Since $\flat : S^\flat \longrightarrow S$ is étale in codimension one

$$\begin{aligned}\flat^*(K_S + C)|_{C'} = (K_{S^\flat} + C')|_{C'} &= K_{C'} \\ &= c^*(K_C) + \sum (r_i - 1) q_i \\ &= c^*(K_C + \sum \frac{r_i - 1}{r_i} p_i).\end{aligned}$$

Thus there is a canonical identification of $\mathbb{Q}$-Weil divisors

$$(K_S + C)|_C = K_C + \sum \frac{r_i - 1}{r_i} p_i.$$

## §5 Log Deformation theory

As indicated in the introduction, the proof of (1.3) involves deforming a rational curve, while maintaining some prescribed contact with a divisor. Here we begin by developing the necessary deformation theory, (5.1-6), a straightforward extension of the results of [19]. Then we will draw a number of corollaries, all of which we believe are of some independent interest.

In the applications, we will often be working on a bug-eyed cover, a non-singular, but (usually) non-separated algebraic space. So we work in this level of generality (which in fact does not in any way complicate the treatment).

Below, by a space, we mean a (not necessarily separated) algebraic space. In all of the Homs we consider, the domain is always assumed to be a projective scheme. See [23] for a discussion of Hom in this context.

First a few remarks on pullbacks and notation (all of which are standard). Given a map $f : X \longrightarrow Y$, and a subspace $K \subset Y$, $f^{-1}(K) \subset X$ indicates the scheme-theoretic inverse image (defined by pulling back the defining equations). If $K$ is an effective Cartier divisor, then $f^{-1}(K) \subset X$ is a locally principal subspace (that is its ideal sheaf is locally principal), and if $X$ is integral and $f(X)$ is not a subset of the support of $K$, then $f^{-1}(K) \subset X$ is again a Cartier divisor, which we will often denote by the more conventional notation $f^*K$. In any case we can always pullback the line bundle $\mathcal{O}_Y(K)$. Given two subspaces $H$, $K$ of a space $Y$, $H \subset K$ means that the space $H$ is a subspace of $K$ (as opposed to the weaker condition of set-theoretic inclusion).

**5.1 Definition-Lemma.** *Let $D \subset X$ and $E \subset Y$ be effective Cartier divisors. Assume $X$ is purely one dimensional.*

*There is a space $\operatorname{Hom}(X, Y, D \subset E)$ with the following universal property. A $T$-point $g : T \times X \longrightarrow Y$ of $\operatorname{Hom}(X, Y)$ is a point of $\operatorname{Hom}(X, Y, D \subset E)$ iff $T \times D \subset g^{-1}(E)$.*

*Proof.* Let

$$k : \operatorname{Hom}(X, Y) \times X \longrightarrow Y$$

be the tautological map. Let

$$G = \pi_* k^* \mathcal{O}_Y(E).$$

$G$ is obviously a vector bundle of rank $\deg D$. The composition

$$k^* \mathcal{O}_Y(-E) \longrightarrow \mathcal{O}_{\mathrm{Hom}(X,Y)\times X} \twoheadrightarrow \mathcal{O}_{\mathrm{Hom}(X,Y)\times D}$$

induces a canonical section of $G$. Let

$$\mathrm{Hom}(X, Y, D \subset E) \subset \mathrm{Hom}(X, Y)$$

be the zero locus. The universal property of the zero locus of this section is obvious from the construction. □

Let $\mathcal{D}$ be a collection of effective Cartier divisors $D_1, D_2, \ldots, D_n$ on $X$ and $\mathcal{E}$ a collection of effective Cartier divisors $E_1, E_2, \ldots, E_n$ on $Y$. We let $E = E_1 + \ldots E_n$. We let $\mathfrak{H} = \mathrm{Hom}(X, Y, \mathcal{D} \subset \mathcal{E})$ denote the scheme-theoretic intersection of the $\mathrm{Hom}(X, Y, D_i \subset E_i)$ inside $\mathrm{Hom}(X, Y)$. $\mathfrak{H}$ then inherits an obvious universal property from (5.1).

**5.2 Lemma.** *Suppose $Y$ is smooth and the divisors $E_1, E_2, \ldots, E_n$ have normal crossings. The sequences*

(1) $$0 \longrightarrow T_Y(-\log E) \longrightarrow T_Y \longrightarrow \oplus \mathcal{O}_{E_i}(E_i) \longrightarrow 0$$

*and*

(2) $$0 \longrightarrow \mathcal{O}_{E_1} \longrightarrow T_Y(-\log E)|_{E_1} \longrightarrow T_{E_1}(-\log[E_1 \cap (E - E_1)]) \longrightarrow 0$$

*are exact.*

*Proof.* For (1) see Lecture 2 of [12]. (2) follows from (1). □

Next we extend (1.2) and (1.3) of [19].

**5.3 Proposition.** *Let $X = C$ be a smooth projective curve. $\mathfrak{H} \subset \mathrm{Hom}(C, Y)$ is cut out by $\deg(D)$ equations. The dimension of any irreducible component of $\mathfrak{H}$ at a point $g$ is at least $\chi(g^* T_Y) - \deg(D)$.*

*If furthermore $(Y, E)$ is log smooth, and $D_i = g^{-1}(E_i)$ for all $i$, then the Zariski tangent space at $[g]$ is $H^0(g^* T_Y(-\log E))$.*

*Proof.* We note first that (1.2) and (1.3) of [19] and their proofs hold when the target is not separated. $\mathfrak{H} \subset \mathrm{Hom}(C, Y)$ is cut out by $\deg(D)$ equations by (5.1). The dimension estimate now follows from (1.2) of [19].

Next consider the commutative diagram

$$\begin{array}{ccccccc} 0 & \longrightarrow & H^0(g^*T_Y(-\log E_i)) & \longrightarrow & H^0(g^*T_Y) & \longrightarrow & H^0(g^*\mathcal{O}_{E_i}(E_i)) \\ & & \downarrow & & \downarrow r & & \| \\ 0 & \longrightarrow & H^0((g|_{D_i})^*T_{E_i}) & \longrightarrow & H^0((g|_{D_i})^*T_Y) & \stackrel{z}{\longrightarrow} & H^0((g|_{D_i})^*\mathcal{O}_{E_i}(E_i)). \end{array}$$

(The first row is pulled back from (5.2.1) and remains exact as $E_i$ pulls back to a Cartier divisor). By the universal property, the tangent space to $\mathrm{Hom}(C, Y, D_i \subset E_i)$ is the kernel of the composition $z \circ r$ and thus by the diagram equal to $H^0(g^*T_Y(-\log E_i))$. The tangent space to $\mathfrak{H}$ is the intersection of the tangent spaces of $\mathrm{Hom}(C, Y, D_i \subset E_i)$ and so by (5.2.1) is equal to $H^0(g^*T_Y(-\log E))$. □

Juggling techniques of Mori we obtain the following:

**5.4 Corollary.** *If $X$ is a smooth variety and $-(K_X + D)$ is ample, then $(X, D)$ is log uniruled. If $-K_X - D$ is pseudo-effective, and $X$ is uniruled, then through a general point of $X$ there is a rational curve meeting $D$ at most twice.*

*Proof.* Let $w$ be a zero dimensional scheme on a curve $C$. We say $w$ has **size** $s$, if $s$ is the maximum length of a subscheme $z \subset w$ supported at a single point. Fix a polarisation $H$ and a general point $x$ of $X$. Among rational curves $f : C \longrightarrow X$ through $x$ with $H \cdot C$ minimum, choose one such that $f^*D$ has greatest size $s$.

Suppose the result fails. Then in the pseudo-effective case, $s \leq C \cdot D - 2$, and $s \leq C \cdot D - 1$ in the ample case. Let $z$ be a subscheme of $f^*D$ of length $s$, let $p$ be the support of $z$, and let $t$ be a point mapping to $x$. Note if $-K_X - D$ is pseudo-effective, $C \cdot D \leq -K_X \cdot C$, with strict inequality if $-K_X - D$ is ample. Thus in any case $\mathrm{Hom}(C, X, z \subset D)$ has dimension at least $-K_X \cdot C + n - s \geq n + 2$ where $n$ is the dimension of $X$. Thus $f$ moves, fixing $x$ and $z \subset f^*(D)$, in a two dimensional family, and so its image must move. Thus we can find a projective surface $S$, a smooth projective curve $B$, and a finite map $S \longrightarrow B \times X$ such that the map $h : S \longrightarrow X$ is generically finite, $S \longrightarrow B$ has general fibre $\mathbb{P}^1$, and for some open set $U \subset B$, $h$ gives a $U$-point of $\mathrm{Hom}(C, X, z \subset D, t \subset x)$, and $h_b : \mathbb{P}^1 \longrightarrow X$ is the map $f$. Since $C \cdot H$ is minimal, $\pi_1$ has reduced irreducible fibres, and thus is a smooth $\mathbb{P}^1$-bundle. Let $Z \subset S$ be the closure of $p \times U$. Then $Z$ is a subscheme of $h^*D$ and a section of $\pi_1$. Let $T$ be the closure of $t \times U$. By assumption there is another component $E$ of $h^*D$, necessarily disjoint from $Z \cup T$. $T$ has negative self-intersection (as $h(T) = x$). Thus (since the cone of $S$ is two dimensional), $Z$ and $E$ have positive self intersection, and $Z \cdot E = 0$, a contradiction. □

(5.4) gives some evidence for (1.10), but it is not very useful for proving (1.1) -assuming both the smoothness for $X$ and ampleness of $-(K_X + D)$ is too strong. The next two corollaries, on the other hand, will prove to be key:

**5.5 Corollary.** *Let $g : X = C = \mathbb{P}^1 \longrightarrow Y$ be a morphism, where $Y$ is smooth and the divisors $E_1, E_2, \ldots, E_r$ have normal crossings. Assume $D_i = g^{-1}(E_i)$. The following hold:*

(1) *If $D$ is supported in at most two points, and $g$ is in a component $Z$ of $\mathfrak{H}$ such that $Z \times \mathbb{P}^1 \longrightarrow Y$ is dominating, then $f^*T_Y(-\log E)$ is semipositive for a general point $f \in Z$.*

*If $g^*T_Y(-\log E)$ is semipositive then (near $g$)*

(2) *$\mathfrak{H} \times \{p\} \longrightarrow Y$ is smooth for any $p \notin D$ and*
(3) *$\mathfrak{H} \times \{p\} \longrightarrow E_j$ is smooth for $p \in D_j \setminus \cup_{i \neq j} D_i$*

*Proof.* (1) follows from (5.3) as in the proof of (1.3) of [19]. (2) and (3) are clear from (5.3). □

The following corollary holds with any number of components, but we present it for at most two components, as this simplifies notation and is sufficient for our needs.

**5.6 Corollary.** *Let $g : X = C = \mathbb{P}^1 \longrightarrow Y$ be a morphism, where $Y$ is smooth and the divisors $E_1, E_2$ have normal crossings. Suppose either*

(i) *$g(\mathbb{P}^1) \subset E_1$, $g^{-1}(E_2) = D_2$, $\deg g^*\mathcal{O}_Y(E_1) = \deg(D_1)$, and $g^*T_{E_1}(-\log(E_1 \cap E_2))$ is semipositive. Set*

$$J = \operatorname{Hom}(C, E_1, D_2 \subset E_2 \cap E_1),$$

*or*

(ii) *$g(\mathbb{P}^1) \subset E_1 \cap E_2$, $\deg g^*\mathcal{O}_Y(E_i) = \deg(D_i)$, and $g^*T_{E_1 \cap E_2}$ is semipositive. Set*

$$J = \operatorname{Hom}(C, E_1, D_2 \subset E_2 \cap E_1) \cup \operatorname{Hom}(C, E_2, D_1 \subset E_1 \cap E_2).$$

*Then $\mathfrak{H}$ is l.c.i and reduced at $g$ and $J$ is a nowhere dense subspace. Furthermore (near $g$)*

(1) *$\mathfrak{H} \setminus J \times \{p\} \longrightarrow Y$ is smooth for any $p \notin D$.*
(2) *$\mathfrak{H} \setminus J \times \{p\} \longrightarrow E_1$ is smooth for any $p \in D_1 \setminus D_2 \cap D_1$.*

*Proof.* We will only deal with case (i), case (ii) is very similar (and not used in the proof of (1.1)).

(5.2.1) and (5.2.2) show that $g^*T_Y(-\log E)$, and $g^*T_Y$ are both semipositive. Let $H = \operatorname{Hom}(\mathbb{P}^1, Y)$. Observe that $J = \mathfrak{H} \cap \operatorname{Hom}(\mathbb{P}^1, E_1) \subset H$. (1-2) are immediate from (5.5.2-3). By (5.3), $H$ and $J$ are smooth at $g$ of dimension $\chi(g^*T_Y)$, and $\chi(g^*T_{E_1}(-\log(E_2 \cap E_1)))$ respectively. By (5.3) every component of $\mathfrak{H}$ has dimension at least $\chi(g^*T_Y) - \deg(g^*E) = \dim(J) + 1$. Thus $J$ is nowhere dense. So by (1-2), $\mathfrak{H} \subset H$ has pure codimension $\deg(D)$. Since it is cut out by $\deg(D)$ equations, it is a complete intersection. In particular it is Cohen-Macaulay, and so since it is generically reduced, it is reduced. □

**5.7 Examples.** As a simple example of (5.6.i), let $g$ embed $\mathbb{P}^1$ as a smooth conic $E = E_1$ in $\mathbb{P}^2$. Let $D = 4[p]$, for a fixed $p \in \mathbb{P}^1$. (5.6.1) says we can deform $g$ in a four dimensional family, while maintaining fourth order contact at $p$. Thus we have the elementary observation that given any conic in $\mathbb{P}^2$ there is a two dimensional family of conics which meets the given conic once.

As another simple example of (5.6.i), suppose $B \subset S^0$ is a log canonical, reduced, but not irreducible, member of $|-K_S|$ (here, as always, $S$ is a rank one log del Pezzo). Then adjunction shows that every irreducible component $E_1$ of $B$, is a smooth rational curve, and $E_1$ meets $E_2 = B - E_1$ normally at two points, say $p$ and $q$. Let $g : \mathbb{P}^1 \longrightarrow E_1$ be the identity. $g^*T_{E_1}(-\log E_1 \cap E_2)$ is trivial by adjunction, so we may apply (5.6.i), with $D_1 = (E_1^2)p$, $D_2 = p + q$, to conclude that $S^0$ is dominated by rational curves (in fact deformations of $E_1$) meeting $B$ twice. In this case $p$ will be a basepoint of the dominating family. One can obtain the same result by counting dimensions in $|E_1|$, since any deformation of $E_1$ is again smooth rational.

We will frequently use the following consequence of (5.5) (sometimes without reference):

**5.8 Corollary.** *Let $B \subset X$ be a reduced divisor on a normal algebraic space. Assume $X$ is a separated variety in a neighbourhood of* $\text{Sing}(B) \cup \text{Sing}(X)$. *Let $V \subset Z$ be a closed subset of codimension at least two, disjoint from* $\text{Sing}(B) \cup \text{Sing}(X)$.

*If $X$ is dominated by a family of rational curves meeting $B \cup \text{Sing}(X)$ at most $d$ times, with $d \leq 2$, then it is dominated by rational curves disjoint from $V$, meeting $B \cup \text{Sing}(X)$ at most $d$ times, and algebraically equivalent to members of the original family.*

*Proof.* After passing to a log resolution of $(X, B)$, replacing $B$ by its total transform and the dominating family by the strict transform, we may assume $(X, B)$ is log smooth. By (5.5.1-3), the general member of the family deforms, maintaining the number of points of contact with $B$, with no basepoints outside of $\text{Sing}(B)$. □

Next a method of proving $X^0$ is uniruled. We use the following:

**Definition.** We say a pair $(X, A)$ of a reduced divisor on a normal variety has **quotient singularities** if there is a quasi-finite surjection $q : U \longrightarrow X$, étale in codimension one, with $U$ and $q^{-1}(A)$ smooth.

**5.9 Corollary.** *Let the pair $(X, A)$ consist of a normal variety and a divisor $A$. Suppose there is a dominating family of rational curves meeting* $\text{Sing}(X) \cup A$ *at most $d$ times, for $d \leq 2$. Suppose, either*

(1) *$(X, A)$ has quotient singularities, or*
(2) *$X$ has terminal, LCIQ singularities (see [23]), and $A$ is empty.*

*Then $X^0$ is dominated by rational curves meeting $A$ at most $d$ times. In particular $X^0$ is uniruled.*

*Proof.* By (4.0), as in the first paragraph of the proof of (4.4), there is a bug-eyed cover $\flat : X^\flat \longrightarrow X$ such that in case (2), $X^\flat$ is l.c.i, and in case (1), $X^\flat$ and $A' = \flat^{-1}(A)$ are non-singular. By (4.9.3), $X^\flat$ is dominated by rational curves meeting $A'$ at most twice.

For (1), apply (5.8) to $(X^\flat, A')$ with $V = \flat^{-1}(\mathrm{Sing}(X))$.

In case (2), let $n$ be the dimension of $X$. Let $C = \mathbb{P}^1$ and $C \longrightarrow X^\flat$ the general map in a dominating family, with $f : C \longrightarrow X$ the induced map. By (2.10) of [23], $\mathrm{Hom}(C, X)$ has dimension at least $-f(C) \cdot K_X + n$ at $[f]$. Let $[f] \in H \subset \mathrm{Hom}(C, X)$ be an irreducible component of that dimension. Let $Y \longrightarrow X$ be a resolution. We may lift $f$ to $f' : C \longrightarrow Y$. Let $H'$ be the irreducible component of $\mathrm{Hom}(C, Y)$ containing $f'$. Since maps in $H'$ dominate $Y$, $H'$ has dimension $-f'(C) \cdot K_Y + n$, by (5.3) and (5.5). Since $H' \longrightarrow H$ is dominant, $f'(C) \cdot K_Y \le f(C) \cdot K_X$. Now by the definition of terminal, it follows that $K_Y \cdot f'(C) = K_X \cdot f(C)$ and $f(C) \subset X^0$. □

*Remark.* For cases where (5.9) applies, we note that a log terminal pair $(S, B)$ of an irreducible curve on a surface has quotient singularities, see Appendix L, and that terminal three-fold singularities are LCIQ, see [23].

For surfaces, we have the following combination of (5.8-9):

**5.9.3 Corollary.** *Let $D \subset S$ be a reduced curve on a log terminal surface. Let $V$ be any finite set of points at which $(S, D)$ is purely log terminal (or equivalently, has quotient singularities). If $S$ is dominated by rational curves meeting $D \cup \mathrm{Sing}(S)$ at most $d$ times, with $d \le 2$, then $S$ is dominated by rational curves disjoint from $V$, and meeting $D \cup \mathrm{Sing}(S)$ at most $d$ times.*

*Proof.* Let $C$ be a general member of a dominating family of rational curves meeting $D \cup \mathrm{Sing}(S)$ at most $d$ times. After blowing up points away from $V$, taking the strict transform of $C$, and then restricting to a neighbourhood of $C$, we can assume $(S, D)$ has quotient singularities. Now apply (5.9). □

**5.10 Corollary.** *The smooth locus of a $\mathbb{Q}$-factorial projective toric variety is uniruled.*

*Proof.* $X$ has quotient singularities and $X^0$ is obviously dominated by copies of $\mathbb{A}^1_*$. Now apply (5.9.1). □

The existence of dominating families of rational curves has implications for the Kodaira dimension:

**5.11 Lemma.** *Let $Y$ be a variety. Suppose that $K_Y + A$ is semi log canonical and that $C$ is the general member of a covering family of rational curves. Let $d$ be the number of times that*

*$C$ meets the union of $A$ with the locus of points where $Y$ is not canonical.*

(1) *If $d \leq 2$ then $(K_Y + A) \cdot C \leq 0$. In particular $K_Y + A$ is not big.*

(2) *If $d \leq 1$, then $(K_Y + A) \cdot C < 0$. In particular the Kodaira dimension of $K_Y + A$ is non-negative.*

*Proof.* Passing to the normalisation of $Y$, we may assume that $Y$ is log canonical. Let $\pi : Z \longrightarrow Y$ be a log resolution of the pair $(Y, A)$. By the definitions of canonical and log canonical, we may write

$$K_Y + B = \pi^*(K_Z + A) + E$$

where $E$ is effective and exceptional, $B$ is reduced, and $\pi(B)$ is contained in the union of $A$ and the locus where $Y$ is not canonical. Thus the strict transform of $C$ meets $B$ at most $d$ times. Replacing the pair $(Y, A)$ by the pair $(Z, B)$ we may assume that $(Y, A)$ is log smooth. Let $f : \mathbb{P}^1 \longrightarrow Y$ be the normalisation of $C$. By (5.5.1), after possibly replacing $f$ by a deformation (with the same contact with $A$), $f^*T_Y(-\log A)$ is semipositive. This proves (1).

Now suppose $d = 1$ (the case $d = 0$ is easier and well known). Let $a = \deg f^{-1}(A)$. By assumption, there is a component $Z$ of $\mathrm{Hom}(\mathbb{P}^1, X, ap \subset A)$ which contains $f$ whose universal family dominates $Y$. The universal family has one dimensional fibres over $Z$, but it has at least two dimensional fibres over $Y$, since $\mathbb{A}^1$ has a two dimensional family of automorphisms. Thus the Zariski tangent space at $[f]$ has dimension at least $\dim(Y)+1$. (2) follows easily by Riemann Roch and (5.3). $\square$

*5.12 Remark.* Note by (5.9.1), if the smooth locus of a log terminal surface $S$ is dominated by images of $\mathbb{A}^1_*$ then in fact the smooth locus is uniruled, that is $S$ is dominated by complete rational curves that miss the singularities entirely.

On the other hand, it is easy to construct rational surfaces with quotient singularities such that $K_S$ is ample (for example the surface in (19.3.2.2.1) with $k \geq 10$). Obviously these surfaces are uniruled, but by (5.11) any dominating family of rational curves meets the singular locus at least three times. Similarly, the bug-eyed covers of these surfaces cannot be dominated by images of $\mathbb{P}^1$, for example by (5.11) and the proof of (5.9).

## §6 Criteria for Log Uniruledness

**Brief Introduction.**

In this section we develop our main techniques for proving log uniruledness. First we will prove (1.3) in the case when the boundary $D$ is non-empty. Thus for (1.3) we are reduced to showing that a rank one log del Pezzo has uniruled smooth locus. In this section we provide two sorts of criteria.

First we will prove the general implication

$$\text{existence of a tiger} \Rightarrow S^0 \text{ uniruled.}$$

As indicated in the introduction, this will complete the first half of the proof of (1.3), and proves (1.3) in all but a bounded collection of cases.

Secondly (6.5) and (6.6) give criteria for when a multiple of a rational curve $Z \subset S$ deforms (as a rational curve) to dominate $S^0$. We will finish the proof of (1.3) in §8-18 by generating a finite set $\mathfrak{F}$ of (families of) $S$, containing any simply connected $S$ without tiger, and then for each $S \in \mathfrak{F}$, finding a rational curve $Z \subset S$ to which we can apply one of (6.5) or (6.6).

At the end of this section we give a number of examples. (6.8) shows that our main criterion, (6.5), is in some sense optimal. (6.10) contains a detailed example application of (6.5). In (6.11) we find the minimal (degree) dominating family of rational curves in $S^0$ in an interesting specific case. In (6.9) we give some examples to convince the reader that one cannot prove (1.1) by (at least a direct naive application of) the *classical* method of constructing rational curves, Mori's bend and break technique (we worry the reader might duck the rather involved analysis ahead if they believed there was a promising alternative).

**Existence of a tiger implies uniruledness.**

The main result of this subsection is the following:

**6.1 Proposition.** *Let $S$ be a normal surface and $\alpha$ an effective $\mathbb{Q}$-Weil divisor on $S$, such that $K_S + \alpha$ is numerically trivial, and not klt. Let $B$ be the support of $\llcorner \alpha \lrcorner$. Then $S$ is dominated by rational curves meeting $B \cup \operatorname{Sing}(S)$ at most twice.*

Note that when $S$ in (6.1) is log terminal, then, by (5.12), the conclusion implies in particular that $S^0$ is uniruled. Thus (6.1) implies (1.14) of the introduction.

We will obtain a stronger form of (6.1) in §20 as a corollary of (1.1), see (1.4).

The main point for (6.1) is the following, the non-empty boundary case of (1.3).

**6.2 Lemma.** *Let $S$ be a rank one log del Pezzo surface. If $B$ is a non-empty reduced curve, such that $K_S + B$ is log terminal and $-(K_S + B)$ is ample, then $(S, B)$ is log-uniruled.*

*Proof.* Note by adjunction and the classification of lt singularities, $B$ has at most two components, and if $B$ is reducible, the two components meet exactly once, normally at a smooth point of $S$.

Let $\flat : S^\flat \longrightarrow S$ be the bug-eyed cover of $S$, (4.5). By (4.14.1) the pair $(S^\flat, B' = \flat^{-1}(B))$ is log smooth.

Let $C$ be a component of $B$ and let $C'$ be the corresponding component of $B'$. By (4.8) and (4.14), $C'$ is non-singular, $K_{C'}$ has negative degree, $C = \mathbb{P}^1$ and there is a surjection $g : \mathbb{P}^1 \longrightarrow C'$. Note $C^2 > 0$ since $S$ has Picard number one. There are two cases.

If $B = C$ has one component we can apply (i) of (5.6) with $E = B'$ and $D$ equal to $\deg(g^*B')p$ for $p$ a general point of $\mathbb{P}^1$. By (5.6.1) $g$ deforms to give a dominating family of maps $h : \mathbb{P}^1 \longrightarrow S^\flat$ such that

$$h^{-1}(B') = (\flat \circ h)^{-1}(B) \supset (\deg(\flat \circ h)^{-1}(B))p.$$

Since the two Cartier divisors above have the same degree, we have equality, and thus $h(\mathbb{P}^1)$ meets $B'$ exactly once, at $h(p)$. By (5.6.1) the general member misses any codimension two subset, in particular $\flat^{-1}(\mathrm{Sing}(S))$, and thus we obtain a family of rational curves dominating $S^0$ and meeting $B$ exactly once.

Otherwise $B$ has exactly two components. Let $B = B_1 + B_2$, where $C = B_1$. $B_2$ meets $C$ normally at a single point, $x \in S^0$, and there is (by adjunction) at most one singular point, $y$, of $S$ along $C$. By (4.9.3), we can assume $\flat \circ g$ is an endomorphism of $\mathbb{P}^1$ totally ramified at $x$ and $y$. Let $p$ be the unique point in $g^{-1}(x)$. Now we apply (5.6.i) with

$$E_1 = B_1', \quad E_2 = B_2', \quad D_1 = \deg(g^*B_1')p, \quad D_2 = \deg(g^*B_2')p = g^{-1}(B_2').$$

The existence of a dominating family follows as before. □

The next Lemma reduces (6.1) to a question about rank one Gorenstein log del Pezzo surfaces, using (6.2) and some standard applications of the MMP. First a little notation. In the notation of (6.1), let $\beta$ be the largest boundary, contained inside $\alpha$. Note that $\alpha$ and $\beta$ have the same support, and agree on the complement of $B$.

**6.3 Lemma.** *To prove (6.1) we may assume,*

1. *There is no morphism $\pi : S \longrightarrow \Sigma$, with general fibre $\mathbb{P}^1$,*
2. *$K_S + \beta$ is log canonical,*
3. *$B$ is non-empty,*
4. *$S$ is a rank one log del Pezzo surface,*
5. *$K_S + B$ is log terminal at singular points,*
6. *$B = \alpha$ (or equivalently, $K_S + B$ is numerically trivial),*
7. *$B \in |-K_S|$, $B \subset S^0$ and $S$ is Gorenstein,*
8. *$S^0$ is simply connected, and*
9. *$B$ is irreducible.*

*Proof.* Suppose there is a morphism $\pi : S \longrightarrow \Sigma$, with general fibre, $F$, isomorphic to $\mathbb{P}^1$. Then $F$ lies in the smooth locus and since $(K_S + B)$ is non-positive on $F$, $F$ meets $B$ at most twice. Hence we have found our dominating family of rational curves. Thus we may assume (1).

Let $f : T \longrightarrow S$ be a log terminal model of the pair $(S, \beta)$, see (17.10) of [27]. By definition of a log terminal model we may write the log pullback (see (1.16) ) as

$$K_T + \tilde{\beta} + E = f^*(K_T + \beta),$$

where $\llcorner E \lrcorner$ is the full reduced exceptional locus, $\tilde{\beta}$ is the strict transform of $\beta$, and $K_T+\tilde{\beta}+\llcorner E \lrcorner$ is log terminal. As discrepancies are not effected by log pullback, we are free to replace $(S,\alpha)$, by $(T, E+\tilde{\beta}+f^*(\alpha-\beta))$. Now $K_S+\beta$ is log terminal, in particular (2) holds.

Suppose $B$ is empty. Then $\alpha$ and $\beta$ are equal, whilst $K_S+\beta$ is klt, a contradiction. Hence (3).

Suppose $f : S \longrightarrow S'$ is any contraction. Let $B'$ be the one dimensional part of $f(B)$. By (5.8), applied with $V = f(B) \setminus B'$, it is enough to find a dominating family of rational curves $C$ which meets $B' \cup \mathrm{Sing}(S')$ at most twice.

The Kodaira dimension of $K_S+(\beta-B)$ is negative, so by (1), the MMP with respect to this divisor gives a birational contraction $f : S \longrightarrow S'$ to a rank one log del Pezzo with $K_{S'}+f_*(\beta-B)$ log terminal and anti-ample. Not every component of $B$ can be contracted by $f$, for otherwise (since $\alpha-\beta$ is supported on $B$) $K_S+\alpha$ is the log pullback of $K_{S'}+f_*(\beta-B)$, contradiction. Replacing $(S,\alpha)$ by $(S', f_*(\alpha))$ we have (1-4).

Now suppose $B$ is not log terminal at singular points. Then we may extract a divisor of the minimal desingularisation of coefficient one (for $K_S+B$) $f : T \longrightarrow S$ and contract the $K_T$-negative ray $\pi : T \longrightarrow S'$ (replacing the boundaries by pullback and pushforward as above). This process improves the singularities so eventually terminates, with (1-5).

Now suppose $B \neq \alpha$. Then $K_S+B$ is anti-ample, and so by adjunction and (5), $K_S+B$ is log terminal. Thus we may apply (6.2). Hence (1-6).

There is a connected cover, $h : T \longrightarrow S$ étale in codimension one, such that $K_T+B' = h^*(K_S+B)$ is trivial (in particular Cartier). By (20.4) of [27], and (5), $K_T+B'$ is log canonical, and log terminal at singular points. It follows, by the classification of log terminal pairs, that $B' \subset T^0$. Thus $T$ is a Gorenstein log del Pezzo. Taking a further cover we may assume, by (3.5), that $T^0$ has trivial algebraic fundamental group. Running the $K_T$-MMP to return to Picard number one, we obtain (1-8) by (3.3).

Finally, by (5.7) we may assume that $B$ is irreducible. Hence (9). □

We will use the following rather ad hoc result in the proof of (6.1):

**6.4 Lemma.** *Let $S$ be a surface with quotient singularities, and let $f : \mathbb{P}^1 \longrightarrow S$ be a generic embedding with image $Z$. Let $B \subset S^0$ be a reduced curve. Assume $Z$ meets $B$ exactly once and* Sing$(S)$ *at most once, and $f^{-1}(B)$ has degree at least two. If $(K_S+B)\cdot Z \leq 0$ then $S^0$ is dominated by rational curves, algebraically equivalent to a multiple of $Z$, meeting $B$ at most twice.*

*Proof.* Let $C = \mathbb{P}^1$. Let $z = Z \cap B$ and if $Z$ meets Sing$(S)$ then let $p = Z \cap \mathrm{Sing}(S)$ and otherwise choose any smooth point $p$ of $Z$ other than $z$. We follow the notation of (4.10). Note that $f : C \longrightarrow Z$, $C' \longrightarrow Z$, and $S^\flat \longrightarrow S$ are all set bijections in a neighbourhood of $z$, and we will also use $z$ to denote the its unique inverse image under any of these maps. Similarly, we'll use the same notation for $B \subset S^0$ and its inverse image on $S^\flat$.

Let $\overline{h} : \mathbb{P}^1 \longrightarrow C$ be of degree $r$ ($r$ as in (4.10.3), which is 1 if $Z \subset S^0$) totally ramified over $z$ and $p$. By (4.7), $\overline{h}$ factors through a map $h : \mathbb{P}^1 \longrightarrow C'$. Let $x = h^{-1}(z)$ and $y = h^{-1}(p)$. Note $h$ is étale at $y$ and totally ramified at $x$. We also indicate by $h$ the induced map $\mathbb{P}^1 \longrightarrow S^\flat$. Note $f^{-1}(B) = s[z]$, $h^{-1}(B) = rs[x]$, for some $s \geq 2$.

Let

$$H = \operatorname{Hom}(\mathbb{P}^1, S^\flat, (rs-1)[x] \subset B).$$

By (5.8) it is enough to show that $H$ gives a dominating family of rational curves (necessarily, by degree considerations, meeting $B$ at most twice). Suppose this fails, or equivalently, that maps in $H$ have fixed image $\flat^{-1}(Z)$. Then by (4.10.2) for

$$J = \operatorname{Hom}(\mathbb{P}^1, C', (rs-1)[x] \subset s[z]),$$

the natural quasi-finite map $J \longrightarrow H$ is surjective near $[h]$.

Obviously, for any non-constant map $\gamma : \mathbb{P}^1 \longrightarrow C'$, the length of $\gamma^{-1}(s[z])$ at any point in its support is divisible by $s$. Thus, since $s \geq 2$, at least set theoretically $J = \operatorname{Hom}(\mathbb{P}^1, C', r[x] \subset [z])$. By (5.3) the latter has Zariski tangent space

$$H^0(h^*T_{C'}(-\log z)) = H^0(T_{\mathbb{P}^1}(-\log x)) = H^0(\mathcal{O}_{\mathbb{P}^1}(1))$$

at $[h]$ and thus has dimension at most two. But by (5.3) the dimension of $H$ is at least $-r(K_S + B)\cdot Z + 3 \geq 3$ at $[h]$, a contradiction, □

Note that (6.4) cannot be extended to the case $Z \cdot B = 1$. For example take for $Z$ a $-1$-curve on a smooth surface, and for $B$ some curve meeting $Z$ normally at one point.

*Proof of (6.1).* By (6.3), we may assume we have properties (6.3.1-9).

Note that $B$ is either a smooth elliptic curve, or a rational curve with a node. If $S = \mathbb{P}^2$ take tangent lines to $B$. Otherwise let $\pi : T \longrightarrow \mathbb{P}^1$ be the $\mathbb{P}^1$-fibration of (3.8), obtained by extracting the $-2$-curve $E$. $B \subset T^0$ is a double cover. By the Hurwitz formula, since $\pi$ has at most one multiple fibre, there is a smooth fibre $F \subset T^0$ meeting $B$ exactly once, with $F$ and $B$ simply tangent (at a point of $B^0$). Let $C \subset S$ be the image of $F$. Since $E$ is a section, $C$ meets $\operatorname{Sing}(S)$ once. Now apply (6.4). □

**Deforming special rational curves $Z \subset S$.**

Here we give some criteria for when a multiple of a rational curve $Z \subset S$ deforms, as a rational curve, to dominate $S^0$. The idea is to lift (an endomorphism of) $Z$ to the bug, and then apply deformation theory.

Here is the main criteria, discussed in the introduction.

**6.5 Lemma.** *Let $S$ be a normal surface with quotient singularities. Let $f : C \longrightarrow S$ be a generic embedding, with $C = \mathbb{P}^1$, such that $f^{-1}(\mathrm{Sing}(S))$ is two points, $p$, $q$. Let $x$, $y$ be the local indices of $Z = f(C)$ at $f(p)$, $f(q)$ (when $f(p) = f(q)$ this means the indices of the two branches). If $-K_S \cdot Z \geq 1/x + 1/y$ then $S^0$ is uniruled, dominated by rational curves algebraically equivalent to a multiple of $Z$.*

*Proof.* Let $u$ and $v$ be the ramification indices of $c : C' \longrightarrow C$ over $p$ and $q$, where $C'$ is the normalisation of $\flat^{-1}(Z)$ . By (4.10.3) $-K_S \cdot C \geq 1/u + 1/v$. Let $h : \mathbb{P}^1 \longrightarrow C$ be a map of degree $uv$, totally ramified above $p$, $q$. By (4.8), there is a lifting $h' : \mathbb{P}^1 \longrightarrow C'$. By (5.3), $\mathrm{Hom}(\mathbb{P}^1, C')$ is smooth at $[h']$ of dimension $u + v + 1$. By (5.3), $\mathrm{Hom}(\mathbb{P}^1, S^\flat)$ has dimension at least $-uvK_S \cdot C + 2$. The result now follows from (4.10.2) and (5.5). $\square$

We note that in the proof of (6.5) we actually obtain a bit more. It's enough to get $K_S \cdot Z \geq 1/r(p) + 1/r(q)$ where $r$ is as in (4.10.3) (note by (4.10.3), $x, y$ divide $r(p)$, $r(q)$, and they are not in general equal). In applications we will know how $\tilde{Z} \subset \tilde{S}$ meets the exceptional locus. From this (and the description of the exceptional locus) it is easy to compute the local analytic index of (some branch of) $Z$. The numbers $r(p)$ and $r(q)$ can also be computed from this description (which after all determines the pair $(S, Z)$ locally analytically), but the analysis is more involved. We use the weaker criterion, and its partner below, as they turn out to suffice.

In most applications of (6.5), $x$ and $y$ will be equal, and each branch of $Z$ will have the same index, the index of the singular point (for example this holds when each branch is log terminal, or by (4.10.3) if neither branch is Cartier and the singularity has prime index). If the inequality in (6.5) fails, we will look for a second curve of the same sort, in order to apply the next result:

**6.6 Lemma.** *Let $S$ be a normal surface with quotient singularities. Let $f : C = \mathbb{P}^1 \longrightarrow S$, $g : D = \mathbb{P}^1 \longrightarrow S$ be two generic embeddings, such that $f(C) \cap \mathrm{Sing}(S) = \{p\}$, $g(D) \cap \mathrm{Sing}(S) = \{q\}$, $Z_1 = f(C) \neq g(D) = Z_2$ and such that $f^{-1}(p)$ and $g^{-1}(q)$ are each a pair of points. Suppose further that each of the local branches of $Z_1$ (resp. $Z_2$) have the same index $n$ (resp. $m$) at $p$ (resp. $q$) where $n$ (resp. $m$) is the index of the surface at $p$ (resp. $q$). Assume there are points $x$ and $y$ of $\mathbb{P}^1$ such that $f(x) = g(y) \in S^0$.*

*Then $S^0$ is uniruled.*

*Proof.* By (6.5) we may assume $nf(C) \cdot K_S = mg(C) \cdot K_S = -1$. As in the proof of (6.5), there are étale maps of degree $n$ and $m$, $h' : \mathbb{P}^1 \longrightarrow C'$ and $j' : \mathbb{P}^1 \longrightarrow D'$, where $C'$ and $D'$ are the normalisations of $\flat^{-1}(Z_1)$ and $\flat^{-1}(Z_2)$. Let $N^\flat$ be the union (with normal crossing) of $C'$ and $D'$ at $h'(x)$ and $j'(y)$. Let $N$ be the union of two copies of $\mathbb{P}^1$ joined along $x$ and $y$. There is an induced morphism $i : N \longrightarrow N^\flat$. Let $\mathcal{N} \longrightarrow B$ be a one dimensional deformation of $N$ to a smooth rational curve. By [19] $\mathrm{Hom}_B(\mathcal{N}, S^\flat \times B)$ has dimension at least 5, while $\mathrm{Hom}(N, N^\flat)$ has dimension four, since a deformation of $i$ is given by étale maps $\mathbb{P}^1 \longrightarrow C^\flat$ and $\mathbb{P}^1 \longrightarrow D^\flat$, with the image of $x$ and $y$ fixed. Thus $S^0$ is uniruled. □

An example application of (6.5) is worked out in (6.10).

We present a three dimensional version of (6.2) (dlt is defined in [27]):

**6.7 Lemma.** *Let $X$ be a projective threefold with a dlt divisor $K_X + E_1 + E_2$ where $E_1$ and $E_2$ are irreducible and normal. Let $C = E_1 \cap E_2$. If $C \cdot E_i$ and $-C \cdot (K_X + E_1 + E_2)$ are both positive then $(X, E_1 + E_2)$ is log uniruled.*

*Proof.* By Adjunction $C = E_1 \cap E_2$ is a smooth rational curve. A local analytic description along $C$ is given in (16.15.2) of [27]. In particular $X$ has quotient singularities along $C$. Let $\flat : X^\flat \longrightarrow X$ be a smooth bug-eyed cover. Then if we set $E_i' = \flat^{-1}(E_i)$ then $E_1'$ and $E_2'$ are smooth and cross normally along $C' = \flat^{-1}(C)$. Adjunction shows that $K_{C'}$ is negative and so there is a surjection $g : \mathbb{P}^1 \longrightarrow C'$. Now apply (5.6), checking semipositivity as in the proof of (6.2). □

**Optimality of (6.5).**

It is possible to have $f : \mathbb{P}^1 \longrightarrow S^\flat$, such that no multiple of $f$ deforms to cover $S^\flat$. Here by a multiple we mean a composition $g = f \circ h$, for $h$ an endomorphism of $\mathbb{P}^1$. Thus it is not sufficient for proving uniruledness to find a rational curve lifting to $S^\flat$, we need some additional local information. For example:

**6.8 Example.** *Suppose $C$ is a rational curve, smooth away from $\operatorname{Sing} S$, meeting $\operatorname{Sing} S$ twice, with each branch of $C$ log terminal, and such that each of the singular points along $C$ has the same index, $m$. Let $f : \mathbb{P}^1 \longrightarrow S^\flat$ be the $m$ to one lift to the bug-eyed cover (as in the proof of (6.5)).*

*If (6.5) fails, that is if $-K_S \cdot C = 1/m$ then no multiple of $f$ deforms to cover $S^\flat$.*

*Proof.* $\flat^{-1}(C)$ has smooth branches, thus $j : C^\flat \longrightarrow S^\flat$ is unramified, and $T_{C^\flat}$ is a subbundle of $j^*(T_{S^\flat})$. $f$ factors through an étale map $\mathbb{P}^1 \longrightarrow C^\flat$ thus $T_{\mathbb{P}^1}$ is a subbundle of $f^*(T_{S^\flat})$. Since $f^*(T_{S^\flat})$ has degree 1, the quotient the quotient is $\mathcal{O}(-1)$ and thus

$$f^*(T_{S^\flat}) = \mathcal{O}(2) \oplus \mathcal{O}(-1).$$

Let $h$ be an endomorphism of $\mathbb{P}^1$ of degree $d$, and let $g = f \circ h$. The Zariski tangent space to $\operatorname{Hom}(\mathbb{P}^1, S^\flat)$ at $g$ is

$$H^0(g^*(T_{S^\flat})) = H^0(\mathcal{O}(2d) \oplus \mathcal{O}(-d))$$

which has dimension $2d + 1$. $\operatorname{Hom}(\mathbb{P}^1, \flat^{-1}(C))$ has dimension at $g$ at least the dimension of $\operatorname{Hom}(\mathbb{P}^1, \mathbb{P}^1)$ at $h$, which is also $2d + 1$. Thus $\operatorname{Hom}(\mathbb{P}^1, S^\flat)$ is smooth at $g$, and equal to $\operatorname{Hom}(\mathbb{P}^1, \flat^{-1}(C))$. □

One instance of (6.8) is the curve $L_{aa'}$ in (15.2).

**6.9 Examples: Trouble with Bend and Break.**

(1) If $C$ is a curve in the smooth locus of $\overline{\mathbb{F}}_e$ then $-K_{\overline{\mathbb{F}}_e} \cdot C \geq e+2$. Note in contrast that if $X$ is a smooth projective $n$-dimensional Fano variety, then the bend and break techniques give that $X$ is covered by rational curves $C$ with $C \cdot -K_X \leq n+1$.

(2) Let $S$ be the cone over a smooth elliptic curve. Then $S$ is a log canonical surface of Picard number one, with $-K_S$ ample, but $S^0$ contains no rational curves. Note that one can choose $C \subset S^0$ any curve, and then apply Mori's Bend and Break argument to produce a rational curve, but no matter how generally one chooses $C$, the rational curve will meet the singular point.

(3) Pick any threefold $Y$, and any smooth curve $C$ in $Y$. Blow up $Y$ along $C$. The exceptional divisor is a ruled surface $\pi : S \longrightarrow C$. Now blow up again, along a section of $\pi$ with very high self-intersection, to obtain a smooth threefold $Z$. Let $E$ be the strict transform of $S$, and $F$ be the general fibre of $E$ over $C$. The cone of $Z$ over $Y$ is two dimensional and $F$ generates one edge. Since $(K_Z + E) \cdot F = -2$, we may, by the contraction theorem, contract $E$ down to $C$ to obtain $X$, a $\mathbb{Q}$-factorial threefold over $Y$. Since $K_Z \cdot F = 0$, $X$ is Gorenstein and canonical. Now the exceptional divisor of $X$ over $Y$ is dominated by $K_X$-negative curves contained in $X^0$, but if $Y$ itself contains no rational curves then every rational curve in $X$ meets $\operatorname{Sing}(X)$.

(4) In [35] Zhang gives examples of log terminal $\mathbb{P}^1$-fibrations $f : T \longrightarrow \mathbb{P}^1$ such that $\kappa(-K_T) = 2$ but $T^0$ is *not* simply connected. In particular $T^0$ is not rationally connected. By (IV.3.10.3) of [26] the fibres of $f$ are the unique family of rational curves dominating $T^0$, while any two points of $T^0$ are connected by a smooth complete $K_T$ negative curve contained in $T^0$ (take a high multiple of an ample divisor).

## 6.10 Example Computation

Here we work through one example application of (6.5) in considerable detail. We hope that after digesting this example, the reader will have little trouble following the analogous, but of necessity rather terse, analysis of §17 and §19, where we apply (6.5) to each of the roughly sixty surfaces $S \in \mathfrak{F}$.

**Choose a configuration in $\mathbb{P}^2$ of a conic $B$, a secant line $A$, and a tangent line $C$. Let $\{a, b\} = A \cap B$ and $c = B \cap C$ (see Figure 3, (8.1)).**

**Let $\tilde{S} \longrightarrow \mathbb{P}^2$ be obtained by blowing up 3 times at $c$, along $C$, (that is further blow ups are at the point of the strict transform of $C$ lying over $c$) 5 times at $b$ along $B$ and 7 times at $a$ along $A$. Let $\tilde{S} \longrightarrow S$ contract the curves on which $K_{\tilde{S}}$ is nonnegative. The resulting surface** has two singular points: $x = (2^C, 7^A, 2, 2, 2, 2)$, index 57, and $y = (2, 2, 4^B, 2, 2, 2, 2, 2, 2)$, index 52, where for example $2^C$ indicates that this $-2$-curve is the strict transform of $C$ (the index can be calculated using (3.1.8) of [27]).

We use Lemma (6.5) to argue that $S^0$ is uniruled. We look for rational curves $Z \subset S$, such that $Z \cap \mathrm{Sing}(S)$ is a single point, $p$, $Z$ has two analytic branches at this point, (we will abuse notation and say $Z$ has a node at $p$, even though either of these branches may be singular, and so the singularity may not actually be a double point) and each branch has the same index, $m$. If $Z$ is such a curve, then (6.5) can only fail if $-K_S \cdot Z = 1/m$ (note $m$ is in this case the global index of $Z$, that is $m$ is the smallest positive integer such that $mZ$ is Cartier).

There are several natural choices for $Z$. We follow the notation of (4.10). Thus $C \longrightarrow Z$ is the normalisation. One could try the strict transform of the tangent line to $B$ at $b$. This curve has a node at $x$, one branch is lt (so index 57), but the other branch is $(2, 7, 2, 2', 2, 2)$, and one checks that the index of this branch is 19. One computes that $-K_S \cdot Z = 2/57$, and so we can't apply (6.5).

Another reasonable choice would be to take $Z$ the strict transform of the secant line from $c$ to $a$. Then $Z$ has a node at $y$, with one lt branch (and so the index is 52) and the other branch

$$(2, 2, 4, 2', 2, 2, 2, 2, 2) = (A_2, 4, 2', A_5).$$

One checks that the index of this non lt branch is 26. One computes $-K_S \cdot Z = 2/52$ so again we can't apply (6.5).

Now take instead for $Z$ the smooth conic which has third order contact with $B$ at $c$ and is tangent to $A$ at $a$. $Z$ is a zero curve on $S$, with two branches at $y$: one is $(A_2, 4, 2, 2', A_4)$ and the other is singular, $(2, 2', 4', A_6)$. By this last notation we mean, $Z$ meets two exceptional divisors of $\tilde{S} \longrightarrow S$, the two marked curves, at their point of intersection. One checks that each of these branches has index 52. Also one computes that

$$-K_S \cdot Z = 2 - 42/52 - 28/52 - 30/52 = 4/52$$

and thus $S$ is log uniruled by (6.5).

**6.11 Rational curves in $S^0(E_8)$.**

Here we work through an interesting and non-trivial example. In fact our proof of (6.1) does not generate the dominating family of minimal $-K_S$-degree. Here we determine such a family for $S(E_8)$. This will not be used in the proof of (1.1).

We follow the notation of (3.6-7). We indicate the two surfaces of (3.7) by $S_N$ and $S_C$, corresponding to a nodal and cuspidal cubic, respectively. Let $N, M \in |-K_{S_N}|$ be the rational curves $N_1$, $N_2$ of (3.6).

The following result should be compared with (6.6).

**6.11.1 Proposition.** *There is a dominating family of rational curves $G \subset S_N^0$ with $G \cdot K_{S_N} = -2$.*

*Proof.* Let $S = S_N$. Let $\mathcal{C} \longrightarrow D$ be a one parameter deformation of a union of two $\mathbb{P}^1$s meeting at a point, to a smooth $\mathbb{P}^1$, with singular fibre $\mathcal{C}_0$. Let $f : \mathcal{C}_0 \longrightarrow S$ be the map given by the normalisations of $N$ and $M$. The dimension of $\mathrm{Hom}(\mathcal{C}_0, S)$ at $[f]$ is four (neither branch can move away from $N \cup M$, thus the node must get sent to $q = N \cap M$, and then we can take any automorphism of either branch fixing the preimage of $q$). But according to (1.2) of [19], $Hom_D(\mathcal{C}, D \times S)$ has dimension at least $\chi(f^*(T_S)) + \dim(D) = 5$. Thus $f$ deforms to a map of a smooth $\mathbb{P}^1$ into $S^0$. □

Fix a flex point $q \in B^0$ to an integral plane cubic $B$. Let $L$ be the flex line. Blowing up 8 times at $q$ along $B$ gives $\tilde{S}(E_8) \longrightarrow \mathbb{P}^2$, see the proof of (3.7). Let $S(B)$ be the corresponding copy of $S(E_8)$. By the construction $B \subset S(B)^0$. We note that if $B_t$ is a flat family of such elliptic curves, the $S(B_t)$ form a flat family. Choose such a family with $B_t$ nodal for $t \neq 0$, and $B_0 = C$ cuspidal. Then there is a family $D_t$ of members of $|-2K_{S_t}|$ with $D_t = M_t + N_t$ for $t \neq 0$, and $D_0 = 2C$.

**6.11.2 Proposition.** *Let $S = S(E_8)$. There is a dominating family of rational curves $J \subset S^0$ with $J \cdot K_S = -2$.*

*Proof.* By (6.11.1) we can assume $S = S_C$. We consider the family $S_t$ defined above. By (6.11.1) we have for each $t \neq 0$ a one dimensional family $\mathcal{G}_t \subset |-2K_{S_t}|$ of rational curves contained in the smooth locus, which includes $N_t + M_t$. After a possible base change, these lie in a family for all $t$. Then the limit, $\mathcal{G}_0 \subset |-2K_{S_C}|$ gives a one dimensional family of rational members, containing $2C$. Since $C \subset S_C^0$, these dominate $S_C^0$. □

Note there are only finitely many rational members of $|-K_{S(E_8)}|$ so the family in (6.11.2) is obviously of minimal degree.

## §7 Reduction to $\pi_1^{\mathrm{alg}}(S^0) = \{1\}$.

Our main goal is to reduce the proof of the empty boundary case in (1.3) to the case when $S^0$ has trivial algebraic fundamental group, (7.2). We also prove, (7.10), the Picard number one, non-empty boundary, case of (1.6).

Additionally, we include some easy lemmas describing how $\pi_1^{\mathrm{alg}}$ is effected by birational transformations, and some implications for $\pi_1$ of the existence of certain rational curves. These will be useful in the hunt analysis. The results are well known, we include proofs for the reader's convenience.

We start with an easy Lemma.

**7.1 Lemma.** *Let $X$ be a $\mathbb{Q}$-Gorenstein Fano variety of dimension $n$.*

*If $K_X^n > n^n$ then $X$ has a special tiger.*

*Proof.* By Riemann-Roch and Serre vanishing, the dimension of the linear system $|-mK_X|$ grows like $m^n K^n/n!$. On the other hand, if $p$ is smooth point of $X$, then it is at most $\binom{mr+n-1}{n}$ conditions on a linear system to have a member of multiplicity at least $mr$ at $p$. So if the given inequality holds, there will be a member $D \in |-mK_X|$ with multiplicity $nm$ at $p$. Then if $E$ is the exceptional divisor of the blow up at $p$, $e(E, K_X + 1/mD) \geq 1$. □

**7.2 Proposition.** *If every rank one log del Pezzo with $\pi_1^{\mathrm{alg}}(S^0) = \{1\}$ is log uniruled then every rank one log del Pezzo is log uniruled.*

*Proof.* Suppose $S$ is any rank one log del Pezzo. If $\pi_1^{\mathrm{alg}}(S^0) \neq \{1\}$, then there is a cover $\pi : S' \longrightarrow S$, where $K_{S'}^2 \geq 2K_S^2$. Running the MMP, we may assume $S'$ is a rank one log del Pezzo, as $K_{S'}^2$ will only increase. Moreover $S$ is log uniruled iff $S'$ is log uniruled.

If this process does not terminate, then eventually $K_S^2 > 4 = 2^2$, and we may apply (7.1) □

We also obtain some easy lemmas describing how $\pi_1^{\mathrm{alg}}$ is affected by the hunt.

**7.3 Lemma.** *Let $g : T \longrightarrow S$ be a birational map between normal surfaces, with exceptional locus $E$. There is a natural surjection $\pi_1^{\mathrm{alg}}(S^0) \longrightarrow \pi_1^{\mathrm{alg}}(T^0)$, which is an isomorphism if $g(E) \subset S^0$*

*Proof.* We note that if $U \subset X$ is an open subset of a normal variety $X$, then $\pi_1^{\mathrm{alg}}(U) \longrightarrow \pi_1^{\mathrm{alg}}(X)$ is surjective (since a connected cover cannot become disconnected when restricted to $U$), and if $X$ is smooth and $U^c$ has codimension at least two, $\pi_1^{\mathrm{alg}}(X) = \pi_1^{\mathrm{alg}}(U)$ (by purity of the branch locus). □

**7.3.1 Corollary.** *In the hunt, if $\pi_1^{\mathrm{alg}}(S_i)$ is trivial, and $\pi(\Sigma_{i+1}) \in S_{i+1}^0$ then*

$$\pi_1^{\mathrm{alg}}(T_{i+1}) = \pi_1^{\mathrm{alg}}(S_{i+1}) = \{1\}.$$

**7.4 Lemma.** *Let $Y_1$, $Y_2$ be two complete $\mathbb{Q}$-Cartier divisors in an integral quasi-projective variety $X$ of dimension at least two. If $Y_i|Y_i$ is nef and big, $i = 1, 2$ then $Y_1 \cap Y_2 \neq \emptyset$.*

*Proof.* Cutting by hyperplanes, compactifying and desingularising, we may assume $X$ is a smooth projective surface. The result now follows from the Hodge index Theorem. □

**7.4.1 Corollary.** *Let $Y$ be an effective complete integral Weil, $\mathbb{Q}$-Cartier divisor in a connected integral quasi-projective variety $X$ of dimension at least two. If the normal bundle of $Y$ is big and nef, then $\pi_1^{\mathrm{alg}}(Y) \longrightarrow \pi_1^{\mathrm{alg}}(X)$ is surjective.*

*Proof.* Its enough to show a connected finite étale cover of $X$ restricts to a connected cover of $Y$. Let $f : X' \longrightarrow X$ be such a cover. Then $f^*(Y)$ has big nef normal bundle, and so is connected by (7.4). □

**7.5 Lemma.** *Let $S$ be a normal projective surface of Picard number one. If $S^0$ is uniruled, then it is rationally connected, and $\pi_1(S^0)$ is finite.*

*Proof.* Just copy the proof of (3.4) of [18], to deduce that $S^0$ is rationally connected. Now apply (7.8) below. □

**7.6 Definition.** Let $C$, $X$ be varieties. We say that $X$ is $C$-connected, if given any two general points $x, y \in X$, there is a generically finite map $C \longrightarrow X$ whose image contains $x$ and $y$.

**7.7 Definition.** A homomorphism of groups is called **almost surjective** if the image is a subgroup of finite index. A group is called **almost Abelian** if it has an Abelian subgroup of finite index.

**7.8 Lemma.** *If $X$ and $C$ are quasi-projective, and $X$ is $C$-connected, then there is a map $f : C \longrightarrow X$ such that the induced map $f : \pi_1(C) \longrightarrow \pi_1(X)$ is almost surjective.*

*Proof.* Let $x \in X$ be a general point. By quasi-projectivity of (the connected components of) $\operatorname{Hom}(C, X)$, there is a variety $U$, a dominant map $g : U \times C \longrightarrow X$, and a multi-section $\Sigma \subset U \times C$ (for the first projection) such that $g(\Sigma) = x$. Replacing $U$ by $\Sigma$ (and pulling back), we can assume $\Sigma$ is a section. By [24] a dominant map of varieties induces an almost surjection on fundamental groups. $\pi_1(U \times C)$ is generated by the images of $\pi_1(\Sigma)$ and $\pi_1(C)$ (where the latter is included via any fibre). Clearly the image of $\pi_1(\Sigma) \longrightarrow \pi_1(X)$ is trivial. □

**7.9 Corollary.** *If $X$ is $\mathbb{A}^1$-connected (resp. $\mathbb{A}^1_*$-connected) then $\pi_1(X)$ is finite (resp. almost Abelian).*

**7.10 Proposition.** *Let $(S, B)$ be a log Fano pair, of Picard number one, with $B$ a non-empty curve. Let $U = S \setminus (B \cup \operatorname{Sing}(S))$. Then $U$ is $\mathbb{A}^1_*$-connected, in particular, by (7.9), $\pi_1(U)$ is almost Abelian.*

*Furthermore, if $U$ is not $\mathbb{A}^1$-connected, then $g^*(T_S(-\log B)) = \mathcal{O}(1) \oplus \mathcal{O}$ for $g : \mathbb{P}^1 \longrightarrow C$ the normalisation of the general member of any family of rational curves dominating $S^0$ and meeting $B$ once.*

*Proof.* By (6.2), $S^0$ is dominated by a family of rational curves meeting $B$ exactly once. Let $g : \mathbb{P}^1 \longrightarrow C$ be the normalisation of a general member of any such family. Suppose $U$ is not $\mathbb{A}^1$-connected. We have $g^{-1}(D) = d[q]$ for some point $q \in \mathbb{P}^1$ and some $d > 0$. Consider the dimension, $n$, of $H = \operatorname{Hom}(\mathbb{P}^1, S, d[q] \subset B)$ at $[g]$. By (5.5) and (5.3), $n = \chi(g^*T_S(-\log B))$.

Let $s \in \mathbb{P}^1$ be a general point, and let $x = g(s)$. The subspace

$$H' = \operatorname{Hom}(\mathbb{P}^1, S, d[q] \subset B, s \to x) \subset H$$

(where the additional notation means that $s$ is required to map to $x$) is cut out by 2 equations. By assumption the image of a map in $H'$ is fixed (equal to $C$). Thus $H'$ has dimension one (there is a one dimensional family of maps with fixed image, since in $H'$ the images of $q$ and $s$ are fixed), and so $n \leq 3$. By (5.5), $g^*(T_S(-\log B))$ is semi-positive. By assumption it has positive degree. Thus the bundle is $\mathcal{O}(1) \oplus \mathcal{O}$.

Assume $d \geq 2$. Then $\operatorname{Hom}(\mathbb{P}^1, S, (d-1)[q] \subset B)$ has dimension at least 4 at $[g]$, and the subspace, $H''$, with $s \to x$, dimension at least 2. Thus (since in $H''$ we fix the images of two points, $q, s$) the images of maps in $H''$ is not fixed. By construction they meet $B$ at most twice. Thus $U$ is $\mathbb{A}^1_*$-connected.

The final possibility is that $d = 1$. Let $C_1$, $C_2 \subset S^0$ be two general members of the family. $C_1 \cap C_2 \neq \emptyset$, since the Picard number is one. As in the proof of (6.11.1), we can deform $C_1 + C_2$ to get a dominating family of integral rational curves $D \subset S^0$ with $B \cdot D = 2$, $-K \cdot D = 4$. The reasoning above shows that $D$ deforms with a general point fixed (no condition on the contact with $B$) –if it did not, then the pullback of $T_S(-\log D)$ to the normalisation of $D$ would be $\mathcal{O} \oplus \mathcal{O}(2)$. Since $D \cdot B = 2$, $D \cap U$ gives a connecting family of images of $\mathbb{A}^1_*$. □

Note $U$ in (7.10) is not in general $\mathbb{A}^1$-connected. For example take $(S, B) = (\mathbb{P}^2, L_1 + L_2)$ for two lines $L_1$, $L_2$. $U = \mathbb{A}^1 \times \mathbb{A}^1_*$, which has infinite fundamental group, and thus cannot be $\mathbb{A}^1$-connected by (7.9)

## §8 Flushness and Preparation for the Hunt

**8.0 Introduction.**

We now take up the second half of the proof of (1.3). As outlined in the introduction, the proof amounts to explicitly determining a finite set $\mathfrak{F}$ of $S$, which includes all $S$ with $\pi_1^{\text{alg}}(S^0) = \{1\}$ but no tiger, and for each $S \in \mathfrak{F}$, explicitly exhibiting a family of rational curves dominating the smooth locus. Once $S$ is in hand, the dominating family will be obtained by finding a special rational curve $Z \subset S$ which satisfies the criteria of (6.5) or (6.6). We find $Z$ in each case by a straightforward but rather ad-hoc analysis. Here we consider the problem of finding $\mathfrak{F}$. As noted in the introduction, this is done by analysing possibilities for a series of simplifying birational transformations, which we call the hunt, and which we now describe. In the hunt definition that follows there are a few choices to be made, and a few points that require proof. We will give the precise details in (8.2).

The hunt generates a series of pairs $(S_i, \Delta_i)$ of rank one log del Pezzo surfaces with boundary with the properties that

(1) $-(K_{S_i} + \Delta_i)$ is ample.
(2) If $K_{S_{i+1}} + \Delta_{i+1}$ has a tiger, then so does $K_{S_i} + \Delta_i$.

The process is as follows:

Given $(S_i, \Delta_i)$, we extract an exceptional divisor $E_{i+1}$ of $\tilde{S}_i$ via $f_i : T_{i+1} \longrightarrow S_i$. We will discuss the choice of $E_{i+1}$ below. Let $\Gamma_{i+1}$ be the log pullback of $\Delta_i$ (see §2). $T_{i+1}$ has Picard number two. It admits a (unique) $K$-negative contraction $\pi_{i+1}$. Define $\Gamma' > \Gamma$ so that $K_{T_{i+1}} + \Gamma'$ is $\pi_{i+1}$-trivial (there is some choice in defining $\Gamma'$). Either $\pi_{i+1}$ is a $\mathbb{P}^1$-fibration, or it is a birational contraction $\pi_{i+1} : T_{i+1} \longrightarrow S_{i+1}$ to a rank one log del Pezzo. In the first case we call $T_{i+1}$ a **net** and the process stops. In the second we define $\Delta_{i+1} = \pi_{i+1}(\Gamma')$.

Given $S$ we inductively carry out the above, starting with $(S_0, \Delta_0) = (S, \emptyset)$. To prove (1.3), we assume $S_0$ has no tiger, and $\pi_1^{\mathrm{alg}}(S^0)$ is trivial, and our goal is to give an explicit description of $\tilde{S}$ as a blow up of $\mathbb{P}^2$. Our method is to at some stage determine $S_i$ (or when there is a net, $T_i$) and then recover $S$ by classifying possibilities for the inverse transformation $S_i \dashrightarrow S_0$. The remainder of (8.0) is devoted to a detailed discussion of the techniques which allow us to control the possibilities. Our aim is to give the reader a feeling for how and why these techniques work. These techniques will prove very efficient for classification; we will never need to consider $S_n$ beyond $S_3$, and in fact the vast majority of the analysis will be on $S_1$ and $S_2$.

In the discussion, we will use the following notation, fixed for the remainder of the paper. Let $A = \pi(E_1)$, $S_1 \ni q = \pi(\Sigma_1)$. Let $a$ be the coefficient of $A$ in $\Delta_1$, that is $\Delta_1 = aA$. We let $B \subset S_2$ be the image of $E_2$, and let $b$ be the coefficient of $B$ in $\Delta_2$.

**Choice of $E_i$.**

The hunt is a series of $K$ non-negative blow ups followed by $K$-negative blow downs, with the blow up and the blow down each of relative Picard number one. Such sequences of transformations are frequently studied in the MMP, see for example [34]. The only choice in defining such a sequence of transformations is which divisor $E_{i+1}$ to extract. The hunt is the series of transformations defined by choosing $E_{i+1}$ of maximal coefficient (in $K_{S_i} + \Delta_i$). This choice, natural from the point of view of tigers, turns out to have remarkably strong geometric consequences, and it is these consequences, more than anything, which makes it possible to classify outcomes of the hunt. In order to explain these geometric consequences, we introduce some definitions:

**8.0.1 The Flush Condition.** *Let $X$ be a normal quasi-projective variety and $\Delta = \sum a_i D_i$ a boundary. Let $m = m(\Delta)$ be the minimum of the non-zero $a_i$. If $\Delta$ is empty, we let $m(\Delta) = 1$.*

**8.0.2 Definition.** Let $E$ be any exceptional divisor over $X$. We say that $K_X + \Delta$, and also the pair $(X, \Delta)$ is:

(1) **flush** (resp. **level**) at $E$ if $e(E, K_X + \Delta) < m$ (resp $\leq$).
(2) **flush** (resp. **level**) if $e(E, K_X + \Delta) < m$ (resp. $\leq$) for all exceptional divisors $E$.

We also have local versions: *Let $Z$ be a subset of $X$. We say $K_X + \Delta$ is flush (resp. level) at, or around, $Z$, if there is some neighbourhood $U$ of $Z$, so that $K_U + \Delta|_U$ is flush (resp. level). We say $K_X + \Delta$ is flush away from, or outside of, $Z$ if $K_X + \Delta$ is flush at $X \setminus Z$.*

*8.0.3 Warning-Remark.* Neither (1) nor (2) in (8.0.2) can in general be checked locally, since $m$ can increase when you shrink $X$ (if we throw away all the components of coefficient $m$). Of course (1) can be checked around the center, $P$, of $E$ on $X$, if there is a component of $\Delta$ of coefficient $m$ which meets $P$.

We will show that in the hunt, $(S_1, \Delta_1)$ is flush (an almost immediate consequence of the choice of $E_1$) and that something slightly weaker, but with the same geometric consequences, holds for all hunt stages, see (8.4.5-6).

The flush condition controls both the singularities of $X$ and of $\ulcorner \Delta \urcorner$, in a manner we will now discuss. Proofs are given in (8.3).

**Some geometric consequences.**

Consider first the local situation, a flush pair $(S, \Delta)$, where $p \in S$ is a surface germ. Let $A = \ulcorner \Delta \urcorner$.

**8.0.4 Lemma.** *If $p$ is singular, then $K_S + A$ is log terminal at $p$.*

Thus from the classification of log terminal pairs, recalled in Appendix $L$, $p$ is a cyclic singularity, $A$ is a smooth curve germ and $\tilde{A}$ meets only one exceptional divisor of the minimal desingularisation over $p$, one of the ends of the chain.

We introduce some convenient notation for describing this situation. The notation is fixed throughout the paper:

**8.0.5 Notation for flush pairs at a point.**

We indicate the singularity type of the triple $p \in A \subset S$ by a vector of integers $\alpha = -(E_1^2, E_2^2, \dots, E_n^2)$, where the $E_i$ are the exceptional divisors of $\tilde{S}$ (over $p$) and $\tilde{A}$ meets $E_1$. We will say $(S, \Delta)$, has a singularity of **type** $\alpha$ at $p$. Let $a$ be the coefficient of $A$, that is $\Delta = aA$. Of course $(S, \Delta)$ is uniquely determined (locally analytically) by $\alpha$ and $a$. We note that this vector notation is the standard way of describing a cyclic singularity, the only new feature here is our specification of the *adjacent curve*, that is, the curve meeting $\tilde{A}$.

The flush condition implies much more than log terminality of $A$, the smaller the coefficient $a$ is, the milder the singularity. The precise statement involves a definition:

**8.0.6 Definition (Spectral Value).** We will say a vector of positive integers $\alpha$ (and also a flush pair $(S, \Delta)$) has **spectral value** $k$ if (in the notation of (8.0.5)) the coefficient of $E_1$ (for $K_S$) has the form $k/r$, where $r$ is the index of (the cyclic singularity defined by) $\alpha$.

**8.0.7 Lemma.** *Suppose the pair $(S, \Delta)$ is flush. Then*

(1) *There is a unique exceptional divisor of maximal coefficient (with respect to $K_S + \Delta$). It is the exceptional divisor adjacent to $A$, that is $E_1$ of (8.0.5).*

(2) *If the pair $(S, \Delta)$ has spectral value $k$ at $p$, then the coefficient $e$ of the pair $(S, \Delta)$ at $p$ is at least $k/(k+1)$. In particular $a > k/(k+1)$.*

Singularities with small spectral value are very restricted. For the next lemma, we will call the vector $\beta$ the **suspension** of $\alpha$ if $\beta$ is obtained from $\alpha$ by successively adding a two on the left.

**8.0.8 Lemma.** *If $\beta = (j, \alpha)$ then the difference of the spectral value of $\beta$ and $\alpha$ is $(j-2)r$, where $r$ is the index of $\alpha$. In particular the spectral value is invariant under suspension, and if $\alpha$ has spectral value $k$, then $\alpha$ is the suspension of*

(a) *the empty string, that is $\alpha = (2, \dots, 2)$, when $k = 0$.*

(b) *(3), that is $\alpha = (2, \dots, 3)$ (or (3)), when $k = 1$.*

(c) *(4) or (3, 2), when $k = 2$.*

Note that without the flush assumption, that is if all we know is that $K_S + aA$ is log terminal, then completely the opposite holds: the smaller $a$ is, the more possibilities there are for $p$ and $A$.

If the curve $A$ is singular at $p$, then by (8.0.7), $S$ is smooth at $p$. The size of the coefficients of $\Delta$ will control the singularities of $A$. For simplicity of exposition, suppose $\Delta = aA$ (that is each component of $\Delta$ has the same coefficient).

**8.0.9 Lemma.** *Assume $(S, aA)$ is flush at $p$.*

(1) *If $A$ has multiplicity $m$ then $(m-1)a < 1$.*

(2) *If $A$ has two smooth branches meeting to order $g$, or a unibranch singularity of genus $g$ then $a < \frac{g}{2g-1}$ (for the definition of these singularities see (11.1) and (11.2)).*

(3) *If $a \geq 4/5$ then $A$ has normal crossings at $p$.*

Now consider the global situation, a pair $(S, A)$, of a curve on a surface, such that $(S, aA)$ is flush (as will be the case in the hunt for $(S_1, A)$). The two cases above can be played against each other. If $A$ has any singularities, then these singularities put bounds on $a$ by (8.0.9). These bounds control the singularities of $S$ along $A$, by (8.0.7-8). More concretely, suppose $A$ has a singularity of multiplicity $m$ at some point, and let $k$ be the spectral value of $(S, \Delta)$ at a singular point $q$. Then by (8.0.7) and (8.0.9)

$$k/(k+1) < 1/(m-1).$$

Thus if $m \geq 3$, $k = 0$, so, by (8.0.8), $S$ is Du Val along $A$. Similarly, if $A$ has a double point of arithmetic genus $g$, then

$$k/(k+1) < g/(2g-1)$$

so for example if $g \geq 3$, $k \leq 1$. Thus by (8.0.8) the singularities along $A$ are at worst of type $(2, \ldots, 3)$.

Singularities of $A$ also control to a lesser extend the singularities of $S$ off $A$, since $a$ bounds the coefficient of $S$ at any singular point, and singularities with small coefficient are fairly restricted. For example, we list the possibilities with $e(S) < 3/5$ in (10.1).

Of course we have the same control whenever we have an upper bound on $a$.

**Implications for the hunt.** In the hunt, $-(K_{S_i} + \Delta_i)$ is ample. Thus the (near) flushness of $(\Delta_i, S_i)$ sets up a useful dichotomy:

*The smaller the coefficients of $\Delta_i$, the stronger our control on the pair $(S_i, \Delta_i)$; the larger the coefficients of $\Delta_i$, the closer we are to a tiger.*

As noted above, one way to obtain bounds on the coefficients is from singularities of $\ulcorner \Delta_i \urcorner$. Thus we have the general philosophy: When $A$ or $A + B$ is quite singular, we expect strong control. As an easy example:

**8.0.10 Sample Lemma.** *If, in the hunt, $A$ has a singularity of multiplicity at least 4, then $S_0$ is Du Val. Any Du Val $S$ has a tiger.*

*Proof.* Note by the definition of the hunt $a > e_0 = e(S_0)$. Thus by (8.0.9.1), $e_0 < 1/3$. An easy coefficient calculation shows $S_0$ must be Du Val, see (10.1). The last remark is an easy consequence of Riemann Roch, see (10.4). □

When $\Delta_2$ has two components, the possibilities for $(S_2, A_2 + B_2)$ will divide into a small number of fairly simple geometric configurations, which we now introduce. We define the configurations in general (that is outside the context of the hunt).

*Let $A$ and $B$ be two rational curves on $S$, such that $K_S + A + B$ is lt at any singular point of $S$. We emphasize that either $A$ or $B$ may have singularities (at smooth points of $S$).*

The following figure will hopefully clarify some of the notation:

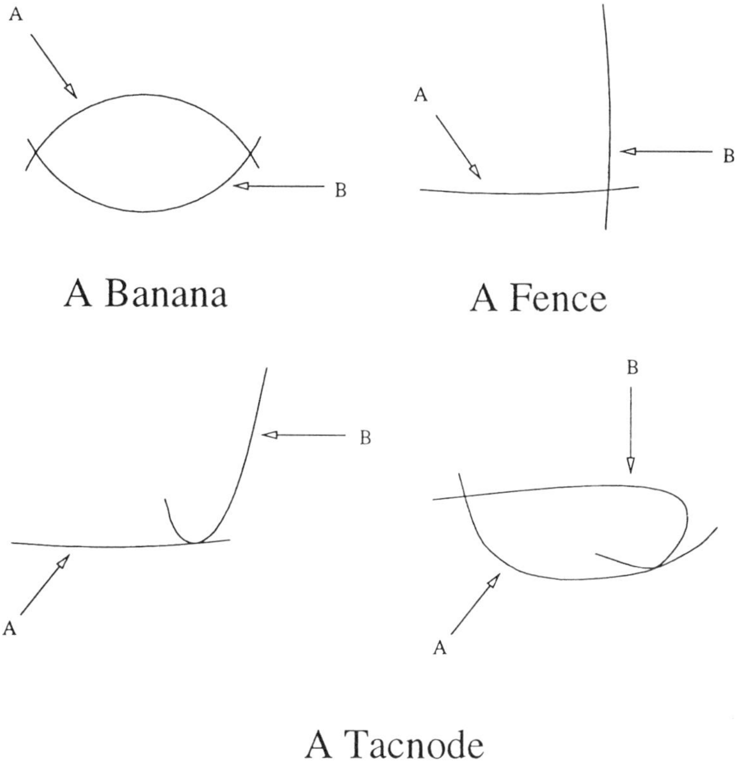


Figure 1

**8.0.11 Definition.** We say that $(S, A+B)$ is a **banana**, if $A$ and $B$ meet in exactly two points, and there normally.

**8.0.12 Definition.** We say that $(S, A+B)$ is a **fence** if $A$ and $B$ meet at exactly one point, and there normally.

**8.0.13 Definition.** We say that $(S, A+B)$ is a **tacnode** if $A \cap B$ is at most two points, there is one point $q \in A \cap B$ such that $A+B$ has a node of genus $g \geq 2$ at $q$ (for the definition, see (11.1)), and if there is a second point in $A \cap B$, then $A$ and $B$ meet there normally.

Another tool which we use repeatedly is a partial classification of the contractions $\pi_i : T_i \longrightarrow S_i$ (locally analytically around $\Sigma_i$). The classification is easy and elementary. We carry it out in §11. With this classification, once we have identified $(S_2, A+B)$ or $(S_1, A)$, we can reverse the process to recover $S_0$.

**8.0.14 Rough Sketch of the proof.**

The general idea of the second half of the proof of (1.3) is as follows (see also the flow chart below). Assuming $S = S_0$ has no tiger (and simply connected smooth locus), we run the hunt. If $A$ is sufficiently singular, we expect, by the above philosophy, very strong control. In fact when $g(A) \geq 2$ (the arithmetic genus) we are able to rule out all but one case, which we can explicitly construct, (15.2). When $g(A) = 1$ or $A$ is smooth we consider the next hunt step. Again we expect strong control if $A_2 + B_2$ has a singularity of arithmetic genus at least two, and in fact we can rule out all possibilities except $(S_2, A_2 + B_2)$ either a smooth Banana, or $(S_2, A_2 + B_2)$ a fence with $B_2$ smooth and $A_2$ of arithmetic genus 1, with a simple node. For such pairs the Bogomolov Bound (9.2) gives very strong global control, and we are able to give a classification, an explicit construction of each possible case, see (13.2) and (13.5). We also have a classification for the inverse transformation $S_2 \dashrightarrow S_0$ (see §11) and so can recover an explicit expression for all possible $S_0$.

We note, in case the reader is puzzled by the occurrence of a node, but not a cusp, in the preceding paragraph, that the triviality of $\pi_1^{\text{alg}}(S^0)$ allows us at various points to simplify the analysis, and in particular to rule out a fence where $A_2$ has a genus 1 cusp. We should also note that, for simplicity of exposition, in the above sketch we have left out a few possibilities. Namely $T_1$ or $T_2$ could be a net, or $A$ might be contracted by $\pi_2$ (so $\Delta_2$ has only one component). Of course in the actual proof we analyse these cases, and as implicitly indicated in the sketch, rule them out.

**8.0.15 Content overview for §8-20.**

The remainder of §8 is divided as follows.

In (8.1) we give a concrete example of the hunt. In (8.2) we make precise the definition of the hunt; there are a few choices to be made, and a few assertions that require proof. In (8.3) we study the general properties of the flush condition. In (8.4) we determine to what degree flushness is preserved by the hunt. Then we give a proposition (8.4.7) which gives a detailed description of the possibilities for the first two hunt steps. The proof of (1.3) then proceeds by an explicit analysis of these possibilities. This is carried out in sections §14-19.

Next we give an overview of the sections, and then a flowchart which shows the logical order of the proof. §10-13 treat a collection of essentially independent technical issues:

In §10 we give a partial classification of $S$ with no tiger, and small coefficient. The results, and notation, are used repeatedly in the hunt analysis. This will be a main tool in the cases $g(A) \geq 2$.

In §11 we give partial classifications for the contractions $\pi_i$, and partial classifications for $\mathbb{P}^1$-bundles. These will be sufficient for classifying the inverse transformations $S_i \longrightarrow S_0$ in the circumstances we will encounter in the hunt analysis.

In §12 we study the linear system $|K_{S_1} + A|$ in the case when $A$ is not smooth. In the cases when $g = 1, 2$ the system turns out to contain a surprising amount of geometric information. For $g = 2$ it defines a $\mathbb{P}^1$-fibration, and for $g = 1$ a birational transformation to a Gorenstein log del Pezzo.

In §13 we give a partial classification of Bananas and Fences. These will be sufficient for classifying $(S_2, A_2 + B_2)$, in the cases we will need to consider.

In §14-19 we analyse the possibilities for the hunt, according to the breakdown of (8.4.7). These sections are essentially independent, and can be read in any order.

Finally in §20 we formally present the proof of (1.1), which will be nothing but a series of references to the preceding sections. This section also includes proofs of the corollaries stated in the introduction.

8.0.16 FINDING $\mathfrak{F}$. ASSUME $S$ HAS NO TIGER AND $\pi_1^{\text{alg}}(S^0) = \{1\}$.
GOAL: FIND $S$ AND SPECIAL RATIONAL CURVE $Z \subset S$

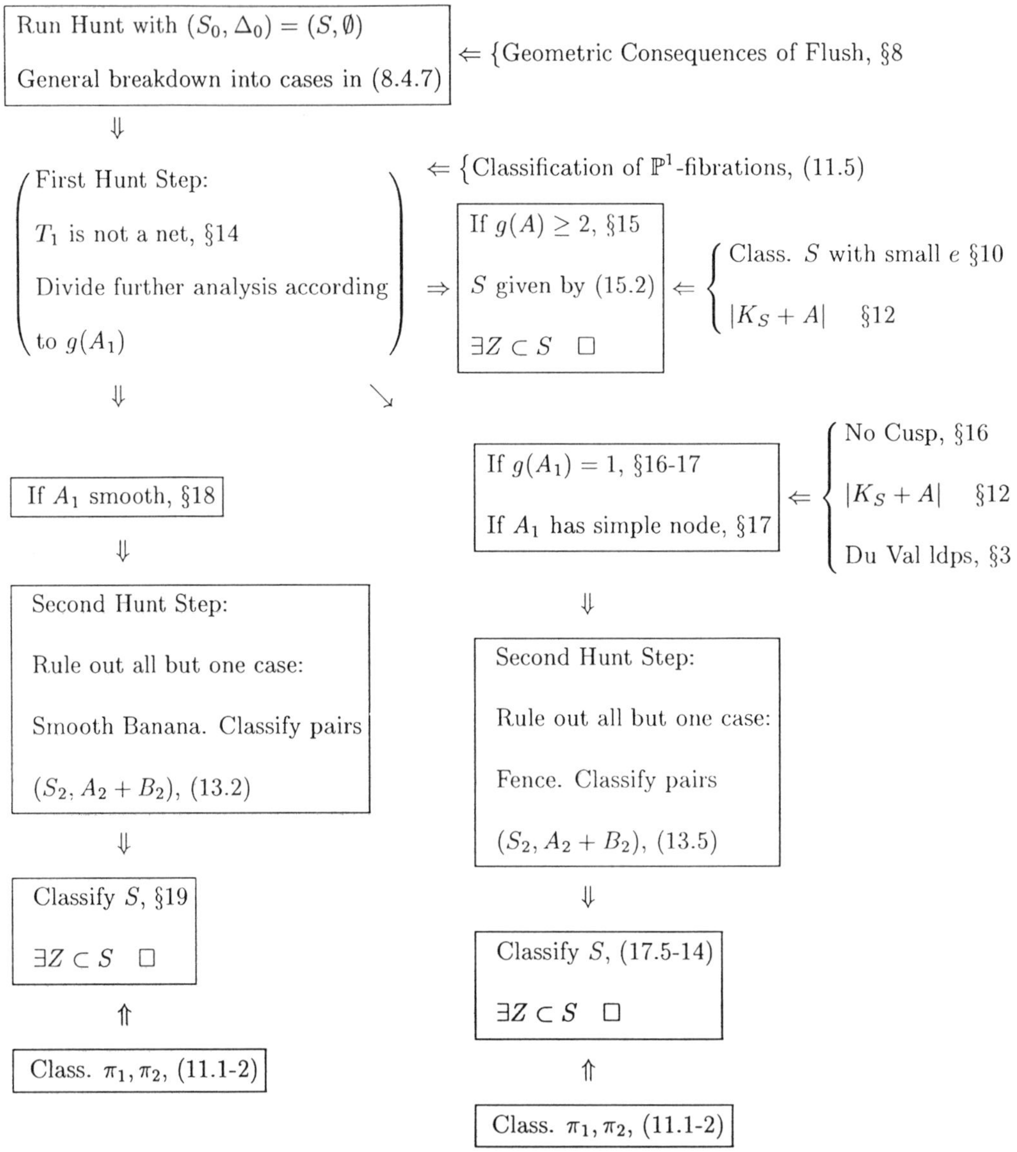


**8.1 An example of the hunt.**

Fix a configuration in $\mathbb{P}^2$ of a tangent line, and a secant line, $D$ and $A$, to a conic $B$. Let $d = D \cap B$, and $\{a, b\} = A \cap B$. Let $h : \tilde{S} \longrightarrow \mathbb{P}^2$ be given by blowing up 3 times at $d$, always along the strict transform of $D$, 5 times at $b$, always along the strict transform of $B$, and 5 times at $a$, always along the strict transform of $A$. Let $\Sigma_1$, $\Sigma_2$, and $\Sigma_3$ be the $-1$-curves of $\tilde{S}$ over $a$, $b$, and $d$ respectively.

Here, as throughout the paper, we will often use the same notation to denote both a curve and its strict transform under some birational transformation.

Let $\tilde{S} \longrightarrow S$ contract the $K_{\tilde{S}}$ non-negative curves. Then $S$ is a rank one log del Pezzo surface, and $\tilde{S}$ is its minimal desingularisation. $S$ has two singular points,

$$x = x_0 = (2^D, 5^A, 2, 2, 2, 2) = (2^D, 5^A, A_4)$$

and

$$y = (A_2, 4^B, A_4)$$

where for example $5^A$ indicates the $-5$-curve of $\tilde{S}$ which is the strict transform of the secant line $A$. $x$ has index 37 and $y$ has index 38. Here is a picture of the dual graph of $\tilde{S}$ over $x$

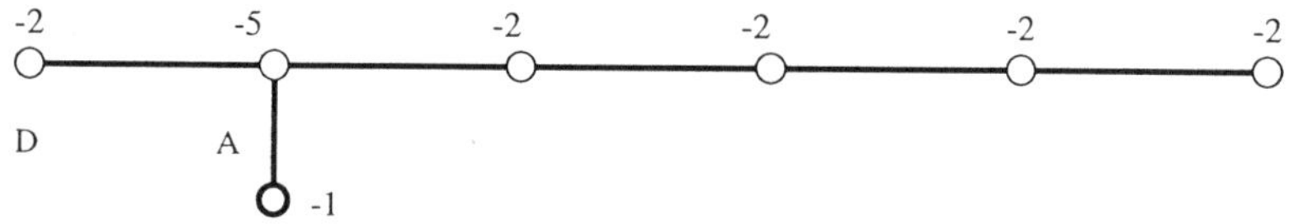


Figure 2

We have chosen the notation to correspond with the outcome of the hunt.

$f_0 : T_1 \longrightarrow S_0 = S$ extracts $E_1 = A$, the $-5$-curve over $x_0$. $\pi_1 : T_1 \longrightarrow S_1$ contracts the $-1$-curve $\Sigma_1$. $A_1 \subset S_1$ is a smooth $\mathbb{P}^1$, $K_{S_1} + A$ is purely log terminal, and there are three singularities of $S_1$ along $A$,

$$(2), (2, 2, 2, 2) \text{ and } (\underline{3}, 2, 2)$$

where the underline indicates the curve which $A$ meets on the minimal desingularisation. A picture of $S$, $\tilde{S}$ and the relevant loci in $\mathbb{P}^2$ is given by:

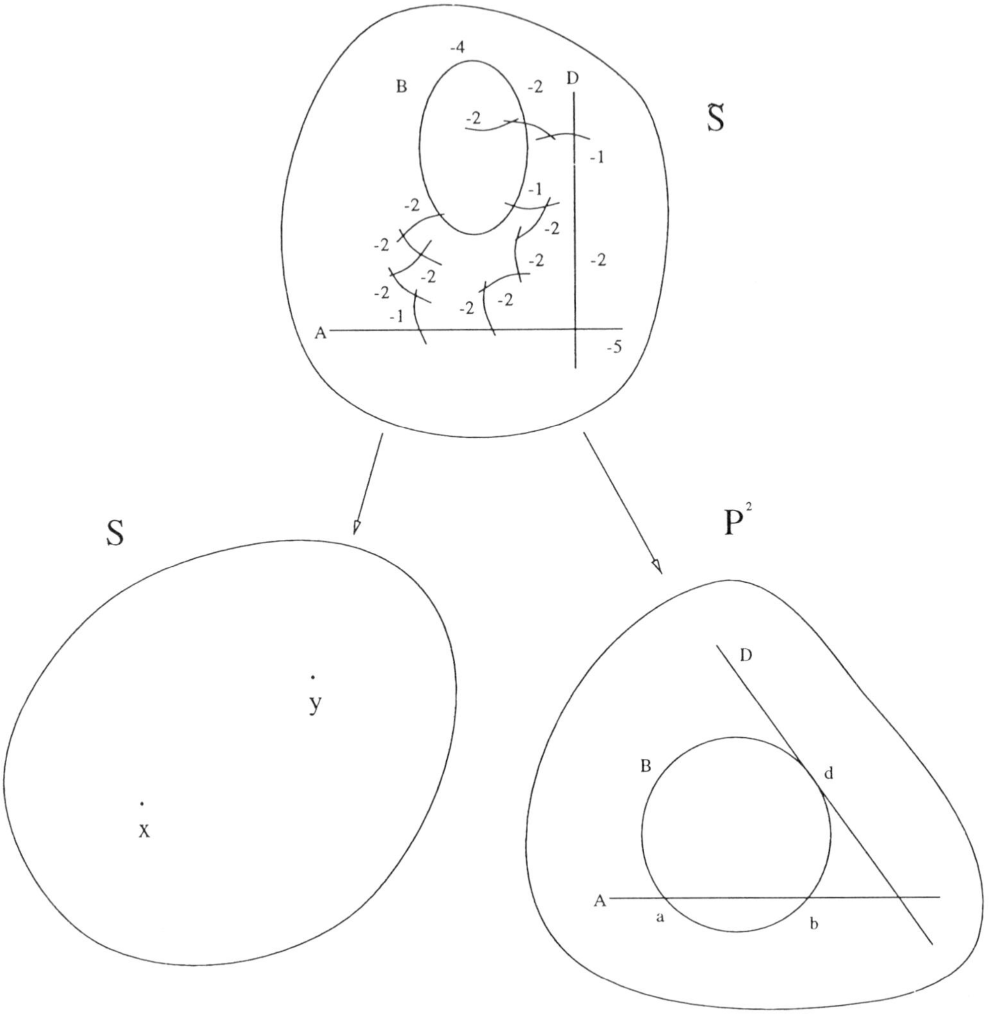


Figure 3

$x_1 \in A$ is the $(3,2,2)$ point, and $f_1 : T_2 \longrightarrow S_1$ extracts the $-3$-curve, $E_2$, which is the strict transform of $B$. $\pi_2$ contracts $\Sigma_2 \subset T_2$. $\Sigma_2$ meets $E_2$ at a smooth point, and meets $A$ at the $A_4$ point (and $A$ and $\Sigma_2$ meet "opposite" ends of the $A_4$ chain).

The configuration $(S_2, A+B)$ is a smooth banana (that is a banana in which both components are smooth). Such configurations are classified in (13.2). $S_2$ is a Du Val surface, with two singular points $A_1 \in A$ and $A_2 \in B$.

$x_2$ is the $A_2$ point along $B$, and $E_3$ is the $-2$-curve meeting $B$. $\pi_3 : T_3 \longrightarrow S_3$, contracts $\Sigma_3$ Let $C$ be the image of $E_3$. $S_3 = \overline{\mathbb{F}}_2$ and $A + B + C \subset S_3$ is a configuration of two sections and a fibre, meeting normally.

*Remark.* If instead on $S_2$ we chose $f_2 : T_3 \longrightarrow S_2$ to blow up the $A_1$ point along $A$, then we would be extracting $E_3 = D$. Again $\Sigma_3$ is contractible, and $\pi_3$ would contract to $\mathbb{P}^2$, and the composition of $\pi_1$, $\pi_2$, and $\pi_3$ would exactly reverse our construction of $S$. However this is not how the hunt proceeds, as one can check by computing coefficients, see (13.2).

We now use the criterion of §6 to prove $S^0$ is uniruled. We need to find rational curves on $S$ with two analytic branches through one of the singular points, and otherwise contained in the smooth locus. For this let $M_b \subset \mathbb{P}^2$ be the tangent line to $B$ at $b$. On $S$, $M_b$ meets $\operatorname{Sing}(S)$ twice, with two analytic branches at $x$:

$$(2', 5, A_4) \text{ and } (2, 5, 2, 2', A_2)$$

where for example $(2, 5, 2, 2', A_2)$ indicates the strict transform of the branch meets only the marked $-2$-curve, and meets it normally. Note the branch with singularity $(2', 5, A_4)$ is thus purely log terminal. Each branch has index 37: since the index of $x$ is prime, this can only fail if the branch is Cartier, and this cannot happen as the two local branches are in fact smooth, see (4.10.3) and (4.12.2). One computes

$$-K_S \cdot M_b = 4/37$$

and thus by the criterion (6.5) $S^0$ is uniruled (and dominated by rational deformations of $37M_b$).

**8.2 General Framework for Hunts.**

Here we make precise the definition of the hunt. A given step has three salient features:

(1) The *underlying transformations* $f : T \longrightarrow S$, $\pi : T \longrightarrow S_1$.

(2) The *scaling*, that is the definition of $\Gamma'$ (for which there is some choice).

(3) The choice of divisor $E$ extracted by $f$.

It will be convenient to have a framework that separates the first two features from the third.

**8.2.4 Scaling.** *Begin with a pair $(S, \Delta)$ of a rank one log del Pezzo surface, and an effective $\mathbb{Q}$-divisor $\Delta$ (we do not assume $\Delta$ is a boundary).*

*Let $f : T \longrightarrow S$ be an extraction of relative Picard number one, of an irreducible divisor $E$ of the minimal desingularisation. Thus $T$ has Picard number two, and so its cone of curves $\overline{NE}_1(T)$ has two edges. One edge is generated by $E$. Let $R$ generate the other edge. Let $x = f(E)$.*

*Let $\Gamma$ be the log pullback of $\Delta$ (for the definition, see (1.16)).*

*We let $\Gamma_\epsilon = \Gamma + \epsilon E$. When $K_S + \Delta$ is numerically trivial, we take $\epsilon = 0$. When $-(K_S + \Delta)$ is ample we assume $0 < \epsilon << 1$, but we will never specify the exact value.*

**8.2.5 Definition-Lemma.** *Assume* $-(K_S + \Delta)$ *is ample (resp. numerically trivial). Then*

(1) $R$ *is* $K_T$*-negative, and contractible. Let* $\pi$ *be the associated map.*
(2) $K_T + \Gamma_\epsilon$ *is anti-ample (resp. numerically trivial).*
(3) $\Gamma_\epsilon$ *is* $E$ *negative (resp. non-positive).*
(4) *There is a unique rational number* $\lambda$ *such that with* $\Gamma' = \lambda\Gamma_\epsilon$, $K_T + \Gamma'$ *is* $R$ *trivial.* $\lambda > 1$ *(resp.* $\lambda = 1$*).*
(5) $K_T + \Gamma'$ *is* $E$ *negative (resp. trivial).*
(6) $\pi$ *is either birational, or a* $\mathbb{P}^1$*-fibration.*
(7) *If* $\pi : T \longrightarrow S_1$ *is birational, and* $\Delta_1 = \pi(\Gamma')$*, then* $K_{S_1} + \Delta_1$ *is anti-ample (resp. numerically trivial) and* $S_1$ *is a rank one log del Pezzo.*
(8) *If* $K_S + \Delta$ *does not have a special tiger, then* $K_S + \Delta$, $K_T + \Gamma'$*, and* $K_{S_1} + \Delta_1$ *are all klt (in particular we have pure boundaries) and none of the three has a special tiger.*

*Proof.* We prove the anti-ample case. The numerically trivial case is similar, but easier. If $K_T \cdot R \geq 0$, then $K_T$ and hence $K_S$ are nef, a contradiction, hence $K_T \cdot R < 0$. $R$ is contractible by the contraction theorem, see [21]. Hence (1). (2) is clear by intersecting with $E$ and $R$, for example by Kleiman's criterion for ampleness. (3) follows from (2), since $K_T \cdot E \geq 0$ by assumption. By (3), $\Gamma_\epsilon$ is $R$ positive (since it's effective), Now (4) follows from (2). (6) is obvious, from (1). (5) follows from (2) and (3), and implies (7), since $S_1$ has Picard number one. (8) is clear from the construction. □

*8.2.6 Remarks.* We introduce $\Gamma_\epsilon$ for technical reasons. In the hunt, if $K_S + \Delta$ is flush, we want $K_T + \Gamma'$ to be flush, but if we scale directly, without introducing $\epsilon$, we can only in general guarantee level. See the proof of (8.4.1). The difference between the geometric implications of flush and level are, we believe, sufficient to justify this additional technicality, see (8.3.6.3).

The process of blowing up, scaling, and blowing down can be carried out in much greater generality, any time one has a log minimal model program. See (2.1) of [22].

**8.2.7 Hunting for a tiger.**

**8.2.8 Definition-Remark.** We will call the transformations $f, \pi$ (with the associated divisors) a **next hunt step** for $(S, \Delta)$ if $e(E, K_S + \Delta)$ is maximal among exceptional divisors of the minimal desingularisation.

Note, there may be several choices of $E$. If $x$ is a cyclic singularity, we allow any choice of maximal $E$, which is not a $-2$-curve (this is always possible by (10.11)) but if $x$ is a non-chain singularity, we require that $E$ be the central curve, which has maximal coefficient by (8.3.9).

The main hunt we will consider is the series of transformations described in the introduction, beginning with $(S_0, \emptyset)$, where we assume $S_0$ does not have a tiger, and our goal is to explicitly determine $S_0$. We will call this **the hunt with** $\Delta_0 = \emptyset$. However occasionally we will run a hunt just to generate a series of (simplifying) birational transformations, see for example the proof of (23.5). The first case will satisfy some additional useful assumptions:

**8.2.9 Definition.** We will say we are **hunting for a tiger** if (in addition to the assumptions of (8.2.4)) the following hold:

(1) $K_S + \Delta$ is klt and anti-ample.

(2) $(S, \Delta)$ is flush at singular points of $S$ lying in the support of $\Delta$.

The hunt with $\Delta_0 = \emptyset$ will indeed satisfy (8.2.9), see (8.4.5-7).

We will use still further notation for the hunt with $\Delta_0 = \emptyset$. Since we have now introduced notation in several places, we will repeat it all here, so the reader will have a convenient reference. It is fixed for the remainder of the paper.

**8.2.10 Glossary of notation for the hunt with $\Delta_0 = \emptyset$.**

$f_i, \pi_{i+1}$ define the next hunt step for $(S_i, \Delta_i)$.

$$S_i \ni x_i = f_i(E_{i+1}), \quad S_{i+1} \ni q_{i+1} = \pi_{i+1}(\Sigma_{i+1}).$$

$\Gamma_{i+1}$ is defined by

$$K_{T_{i+1}} + \Gamma_{i+1} = f^*(K_{S_i} + \Delta_i).$$

$\Gamma_\epsilon = \Gamma_{i+1} + \epsilon E_{i+1}$. $\Gamma' > \Gamma_\epsilon$ is defined in (8.2.5). It is $\pi_{i+1}$-trivial. $\Delta_{i+1} = \pi_{i+1}(\Gamma')$ and satisfies

$$K_{T_{i+1}} + \Gamma' = \pi^*_{i+1}(K_{S_{i+1}} + \Delta_{i+1}).$$

$$S_1 \supset A = \pi_1(E_1), \quad S_2 \supset B = \pi_2(E_2) \quad S_3 \supset C = \pi_3(E_3).$$

If there seems a possibility of confusion, we will use a subscripts, $A_i, B_i, C_i$ to indicate the strict transform of $A, B, C$ on $S_i$.

Let $a, b, c$ be the coefficients of $A, B, C$ in $\Delta_1, \Delta_2, \Delta_3$. These are also the coefficients of $E_1$, $E_2$, $E_3$ in $\Gamma'_1$,$\Gamma'_2$,$\Gamma'_3$. Let $a_i, b_i, c_i$ be the coefficient of $A_i, B_i, C_i$ in $\Delta_i$.

Note that because of the scaling, the coefficients $a_i, b_i, c_i$ each strictly increase with $i$. Flushness and scaling imply $a_i > b_i > c_i$. .

Let $e_i$ be the coefficient of $E_{i+1}$ in $(S_i, \Delta_i)$. This is also the coefficient of the pair $(S_i, \Delta_i)$, see (8.3.2.2) and (8.3.5.4).

We let $e_i$ be the coefficient of $K_{S_i} + \Delta_i$.

We let $E_1$ be a $-k$-curve and $E_2$ a $-j$-curve.

If $\Sigma_i$ passes through a chain singularity $w$ of $T_i$, the notation $(a_1, a_2, \ldots, a_n)$ denotes the negative of the self-intersection of the curves in the minimal resolution of $w$, a prime above (resp. underline) $a_i$ means that $\Sigma_i$ (resp. the strict transform of $\Delta_i$) meets the corresponding curve on $\tilde{T}_i$. We let $\overline{\Sigma}_i$ be the image of $\Sigma_i$ on $S_{i-1}$. We will write $\overline{\Sigma}$ for $\overline{\Sigma}_1$.

**8.3 Properties of the flush condition.**

In this sub-section we have two main goals. We want to obtain the geometric consequences of the flush condition discussed in (8.0), and to determine how the flush condition is effected by the operations of the hunt, namely extracting or contracting a divisor, and scaling a boundary.

**8.3.0 Notation:** *Throughout (8.3) we consider a boundary* $\Delta$*, with support* $D$*.*

We begin with some simple properties that hold in any dimension.

**8.3.1 Lemma.** *Let* $\Delta' \geq \Delta \geq 0$*, with* $\Delta'$ *a boundary, on a normal* $\mathbb{Q}$*-factorial variety* $X$*. Suppose* $f : Y \longrightarrow X$ *is a birational morphism with irreducible divisorial exceptional locus,* $E$*. Assume* $e = e(E, K_X + \Delta) \geq 0$*. Let* $\Gamma$ *be the log pullback of* $\Delta$*, that is* $\Gamma = eE + \tilde{\Delta}$*. Assume* $\Gamma$ *is a boundary.*

(1) *If* $K_X + \Delta$ *is level and* $E$ *has maximal coefficient (for* $(X, \Delta)$*) then* $K_Y + \Gamma$ *is level.*
(2) *If* $K_Y + \Gamma$ *is flush, then* $K_X + \Delta$ *is not flush for* $F$ *iff* $F = E$*, and the coefficient of* $E$ *in* $\Gamma$ *is at least as large as the coefficient of some non exceptional component of* $\Gamma$*.*
(3) *If* $X$ *is terminal and* $K_X + \Delta'$ *is flush then* $K_X + \Delta$ *is flush.*

*Proof.* (1) and (2) are obvious.

We prove (3) by induction on the number of irreducible components in the support of $\Delta$. If this is zero, the result holds by the definition of terminal. If there is more than one component, then dropping a component clearly preserves flushness, since $m$ (of (8.0.1)) will only go up, and the coefficient of any divisor will only go down. Thus we may assume $\Delta'$ and $\Delta$ have the same support, but are not equal. Let $\Theta$ be the boundary $\Delta' - \Delta$, and $\Delta(t)$ the divisor $\Delta + t\Theta$. Let $F$ be any exceptional divisor. Then $f(t) = e(F, K_S + \Delta(t))$ and $g(t) = m(\Delta(t))$, the smallest coefficient of any component of $\Delta(t)$, are both affine functions of $t$. Consider the smallest value of $t$, $t_0$, such that $\Delta(t)$ is a boundary. $t_0 < 0$ since $0 < \Theta$. $\Delta_{t_0}$ has smaller support than $\Delta$, thus by induction, and the definition of flush, we have $g(t_0) > f(t_0)$ and $g(1) > f(1)$. Thus $g(0) > f(0)$, or equivalently, $K_X + \Delta$ is flush at $F$. □

**For the rest of (8.3) we work on the germ of a log terminal surface** $p \in S$**.**

Recall that throughout (8.3), $D = \ulcorner \Delta \urcorner$, the support of $\Delta$.

**8.3.2 Lemma.**

(1) *If* $(S, D)$ *has normal crossings (in particular,* $p$ *is smooth) then* $K_S + \Delta$ *is level, and flush if* $\Delta$ *is a pure boundary.*
(2) *Assume* $(S, D)$ *does not have normal crossings (that is either* $p$ *is singular, or* $D$ *has worse than a simple node). Let* $f : T \longrightarrow S$ *be a log resolution of* $(S, \Delta)$ *(for the definition see §2). Assume* $e(F, K_S + \Delta) \leq 1$ *for any* $f$*-exceptional divisor* $F$*. Then for any exceptional divisor* $V$ *there is an* $f$*-exceptional divisor* $F$ *such that* $e(V, K_S + \Delta) \leq$

*$e(F, K_S + \Delta)$. In particular $K_S + \Delta$ is flush (resp. level) iff it is flush (resp. level) at all exceptional divisors of $f$.*

*Proof.* For (1) it is convenient to prove a little more. We consider only the level case (the flush case is nearly identical). We show that if $(S, A + B)$ has normal crossings, $a, b \leq 1$, where we allow negative numbers, then $e(V, K_S + aA + bB) \leq \min(a, b)$ for any exceptional divisor over $p = A \cap B$.

$V$ can be obtained by a sequence of smooth blow ups. We induct on the number of these blow ups. Let $f : T \longrightarrow S$ be the blow up at $p$, with exceptional divisor $E$. We have

$$K_T + a\tilde{A} + b\tilde{B} + (b + a - 1)E = f^*(K_S + \Delta).$$

Thus $e = e(E, K_S + aA + bB) = (b + a - 1) \leq \min(a, b)$. If $V$ lies on a further blow up, then $V$ is centered over some point $x$ of $E$. Since $A$ and $B$ are separated by $f$, we may assume (by switching notation if necessary) that $x \notin B$. Then by induction

$$\begin{aligned} e(V, K_S + aA + bB) &= e(V, K_T + a\tilde{A} + b\tilde{B} + eE) \\ &= e(V, K_T + a\tilde{A} + eE) \leq e \leq \min(a, b). \end{aligned}$$

For (2). Write

$$K_T + \Gamma = f^*(K_S + \Delta).$$

Here $\Gamma$ is in general only a subboundary (for the definition, see §2). Assume $V$ is centered over some point $x$ on an irreducible component, $F$ of $E$. Let $l$ be the coefficient of $F$ in $\Gamma$. By the proof of (1),

$$e(V, K_S + \Delta) = e(V, K_T + \Gamma) \leq l = e(F, K_S + \Delta) \quad \square$$

**8.3.3 Lemma.** *Assume $p$ is singular. If the coefficient of every exceptional divisor of the minimal resolution $\pi : \tilde{S} \longrightarrow S$ in $K_S + D$ is strictly less than one, then $K_S + D$ is log terminal.*

*Proof.* Suppose $(S, D)$ is not log terminal. $D$ is not empty, since $S$ is log terminal by assumption. Let $E$ be the (reduced) exceptional locus of $\pi$. By (8.3.2.2), $f$ is not a log resolution for $(S, D)$. By the classification of lt singularities, $E$ has normal crossings, thus $D$ does not have normal crossings with $E$. As the computation of coefficients is purely numerical, we may replace $\tilde{D}$ by a disjoint union of analytic discs, each meeting $E$ normally, and such that at least two discs meet the same irreducible component of $E$, without changing the coefficients of any of the exceptional divisors. But then, from the classification of log terminal singularities in Appendix $L$, the pushforward of the new configuration (for which $f$ is a log resolution) is not log terminal, and so by (8.3.2.2), the coefficient of some exceptional divisor must be at least one, a contradiction. $\square$

*8.3.4 Remark.* If one only assumes in (8.3.3) that the coefficients are at most one, one cannot in general conclude that $K + D$ is log canonical at singular points. For example take a cyclic singularity with exactly one exceptional curve, $E$, and assume $\tilde{D}$ is simply tangent to $E$; numerically this is indistinguishable from a log canonical node. However using the same proof, and the classification in Appendix $L$ one can easily classify the exceptions. We note in particular that if $K + D$ is not log canonical at $p$, then there are at most two exceptional divisors in the minimal resolution over $p$.

**8.3.5 Lemma.** *Let $f : T \longrightarrow S$ extract the irreducible divisor $E$. Assume $e(E, K_S + \Delta) \geq 0$ and let $\Gamma$ be the log pullback of $\Delta$. Let $\pi : \tilde{S} \longrightarrow S$ be the minimal desingularisation.*

(1) *If $p$ is singular, and $K_S + \Delta$ is flush at every $\pi$-exceptional divisor $F$ then $K + D$ is lt and $K_S + \Delta$ is flush.*
(2) *If $K_S + D$ is lt and $K_S + \Delta$ is level at every $\pi$-exceptional divisor $F$ then $K_S + s\Delta$ is flush for any $s > 1$ such that $s\Delta$ is a boundary.*
(3) *Suppose $p$ is singular. If $f$ is a $K_T$-contraction and $K_T + \Gamma$ is flush at every exceptional divisor of the minimal desingularisation of $T$, then $K_S + \Delta$ is flush.*
(4) *Suppose $p$ is singular, and $K_S + D$ is log canonical. Then for any exceptional divisor, $V$, there is some $\pi$-exceptional $F$ with $e(V, K_S + \Delta) \leq e(F, K_S + \Delta)$.*

*Proof.* We first prove (1). Suppose the smallest coefficient of $\Delta$ is $\lambda$. Consider the function $f(t) = e(K_S + tD, F)$. Now $f(0) \geq 0$ and as the coefficient of $K_S + \lambda D$ is at most the coefficient of $K_S + \Delta$,

$$f(\lambda) \leq e(F, K_S + \Delta) < \lambda,$$

where the final inequality is from the definition of flush. As $f$ is an affine function, $f(t) < t$, for all $t \geq \lambda$. In particular $e(F, K_S + D) < 1$, so $K_S + D$ is log terminal by (8.3.3). Thus, by the classification of log terminal singularities, Appendix $L$, $\pi$ is a log resolution of $(S, \Delta)$ and so $K_S + \Delta$ is flush by (8.3.2.2). Hence (1).

Now we prove (2). The minimal desingularisation is a log resolution of $K_S + t\Delta$, thus by (8.3.2.2), we need only prove flushness at each $F$. Note $D$ is irreducible, say $\Delta = aD$. Define $f(t)$ as in (1). $f(0) \geq 0$, $f(a) \leq a$, and $f(1) < 1$. Thus $f(t) < t$ for $1 \geq t > a$. Hence (2).

Now we prove (3). By (1) we only need to check flushness at each exceptional divisor, $F$, of $\pi$. Since $E$ is not an exceptional divisor of $\pi$ (it is $K_T$-negative), $F$ is an exceptional divisor of the minimal desingularisation of $T$. Thus

$$m(\Delta) \geq m(\Gamma) > e(F, K_T + \Gamma) = e(F, K_S + \Delta)$$

and $K_T + \Delta$ is flush. Hence (3).

Finally for (4), since $D$ is log canonical, $\pi$ is a log resolution. Thus the result follows from (8.3.2.2). □

Here are a few concrete examples:

**8.3.6 Examples.** Let $p \in C \subset S$ be a curve germ on $S$.

(1) If $C$ has an ordinary cusp at $p$, and $p$ is a smooth point of the surface $S$, then $K_S + aC$ is flush iff $a < 4/5$, and level iff $a \leq 4/5$.

(2) Suppose $K_S + C$ is lt at $p$, and $p$ has type (3) (in the sense of (8.0.5)). Then $K_S + aC$ is flush at $p$ iff $a > 1/2$ and level iff $a \geq 1/2$.

(3) Suppose $K_S + C_1 + C_2$ is lc at $p$, for two irreducible curves $C_1, C_2$. Assume $a, b > 0$. Then $K_S + aC_1 + bC_2$ is flush at $p$ iff $p$ is a smooth point and $\max(a, b) < 1$, and it is level but not flush at $p$ iff either $p$ is smooth and $\max(a, b) = 1$, or $p$ is Du Val and $a = b$, or $a = b = 1$.

*Proof.* For (1). By (8.3.2.2) we need only consider coefficients on a log resolution of $(S, C)$. Such a resolution is obtained by blowing up 3 times over $p$, always on the strict transform of $C$. One computes the coefficient of each exceptional divisor. The determining coefficient is

$$e(\Sigma_3, K_S + aC) = 6a - 4$$

where $\Sigma_3$ is the exceptional divisor of the final blow up (so the unique $-1$-curve over $p$).

For (2). We need only consider $e(E, K_S+aC)$, where $E$ is the $-3$-curve over $p$. One computes $e(E, K_S + aC) = 1/3 + a/3$.

For (3). The flush case follows from (8.3.5.1) and (8.3.2.1). The smooth part of the level case is easy, using (8.3.2.1). So suppose $p$ is singular. By (8.3.5.4) we need only consider coefficients of exceptional divisors from the minimal desingularisation. Let $E$ be such a divisor. Assume $b \geq a$. Let $f(t, s) = e(E, K_S + tC_1 + sC_2)$. $f(0, 0) \geq 0$, and zero iff $p$ is Du Val. $f(1, 1) = 1$. $f(a, b) \leq f(b, b)$ with strict inequality unless $a = b$. By affineness, $f(b, b) \geq b$, with equality iff $p$ is Du Val, or $b = 1$. The result follows. □

*Remark.* (8.3.6.1) shows that when we scale in the hunt, flushness might be lost. It will turn out this can only happen at smooth points of $S$. The other way that flushness can be lost in the hunt, is if at some stage a component of $\Delta_i$ is contracted, (8.3.1.2).

**8.3.7 Lemma.** *Suppose $p$ is smooth and the pair $(S, \Delta)$ is flush. Let $\Delta = \sum a_i D_i$. Then*

(1) *If $M_i$ is the multiplicity of $D_i$ then $\sum a_i M_i - 1 < m$. In particular $m < 1/(M-1)$ where $M$ is the multiplicity of $D$.*

(2) *If $D$ has a node of genus at least two at $p$ (see (11.1)), and the coefficient of the two branches of $B$ at $p$ are $a \geq b$, then $2a + b < 2$.*

(3) *If $p$ is a cusp of $D$ (see (11.2)) and $a$ is the coefficient of the branch of $D$ at $p$, then $a < 4/5$.*

(4) *If $m \geq 4/5$, then $D$ has normal crossings.*

*Proof.* Let $\pi : R \longrightarrow S$ be the blow up of $S$ at $p$. Then the coefficient of the exceptional divisor is $\sum a_i M_i - 1$, which is less than $m$ by assumption. Hence (1).

(2) and (3) may be proved similarly. (4) is immediate from (1-3), since (2-3) are the two possibilities for $D$ of multiplicity two, see (II.8) of [4]. □

*Proof of (8.0.8).* Let $s$ be the index of $\beta$ and $r_2$ the index of the chain obtained by deleting the leftmost integer from $\alpha$. Now $s = jr - r_2$, and the coefficient of the exceptional divisor of $\alpha$ on the left is $(r - r_2 - 1)/r$. The coefficient of $E_1$ is $((j-2)r + r - r_2 - 1)/s$. □

*Proof of (8.0.7).* $K_S + D$ is log terminal by (8.3.5.1). Let $E$ be the divisor of maximal coefficient, and let $\pi : T \longrightarrow S$ extract $E$. By (8.3.5.4) and (8.3.1.1), $K_T + \Gamma = \pi^*(K_S + \Delta)$ is level. From the classification of log terminal singularities, it is enough to show $G = \ulcorner \Gamma \urcorner$ is log terminal at the point $p = E \cap \tilde{D}$. By (8.3.5.1) we need only show $K_T + \Gamma$ is flush at $p$. The coefficient of $\tilde{D}$ in $\Gamma$ is strictly minimal, since $(S, \Delta)$ is flush, thus $K_T + \Gamma$ is flush at $p$ by (8.3.6.3). Hence (1).

For (2). Suppose $p$ has index $r$, and $\Delta = aD$. $K_S + D$ is log terminal, by (8.3.5.1) and so the different of $K_S + D$ is $(r-1)/r$, see (L.2). Now the coefficient of $K_S + aD$ is $(k/r) + a(r-1-k)/r$, which is less than $a$. Thus $a > e > x$, where $x$ is the solution to the recursive equation

$$x = (k + x(r - 1 - k))/r,$$

which is $k/(k+1)$. Hence (2). □

We note that the proof of (8.0.7) also gives

**8.3.8 Lemma.** *(in the notation of (8.0.5))*

$$e(E_1, K + \lambda D) = (k/r) + \lambda(r - 1 - k)/r = \lambda(r-1)/r + (1 - \lambda)(k/r).$$

**8.3.9 Lemma.** *Let $x$ be a non-chain singularity with branches $\beta_1, \beta_2, \beta_3$. Assume the central curve is a $-l$-curve.*

*$e(x)$ is the coefficient of the central curve, and is a rational number of the form $\frac{k}{k+1}$, for some positive integer $k$.*

*Moreover if $\beta_1 = \beta_2 = (2)$, and $\alpha = (l, \beta_3)$ then $k$ is the spectral value of $\alpha$ (see figure 4)*

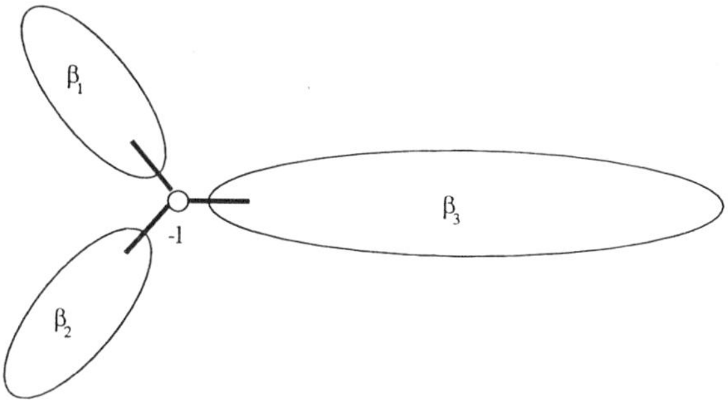


Figure 4

Here ovals denote chains, as in Chapter 3 of [27].

*Proof.* We show first the central divisor has maximal coefficient. Let $f : T \longrightarrow S$ extract a divisor $E$ of the minimal resolution of maximal coefficient, $e$. We assume $E$ is not central, thus it contains at most one non-chain singularity $y$. $K_T + E$ is not log terminal at $y$. By the choice of $E$, $K_T + eE$ is level at exceptional divisors of the minimal resolution. Let $f(t) = e(F, K_T + tE) - t$ for $F$ an exceptional divisor of the minimal resolution over $y$. $f(0) \geq 0$, $f(e) = 0$. Thus $f(1) \leq 0$. Now by (8.3.3), $f(1) = 0$ (for some $F$), thus $f(0) = 0$, and so $y$ is Du Val. Also by (8.3.4), $K_T + E$ is log canonical at $y$. Thus by Appendix $L$ two of the branches at $y$, are (2). Let the third branch be $(2, \ldots, 2) = A_t$ with the central curve meeting the left most $-2$-curve. Let the coefficients of the curves over $y$ be $e_1, \ldots, e_t$. Let $e = e_{t+1}$. Let $e_0$ be the coefficient of the central curve, $a$, $b$ the coefficients of the two $-2$-curves in the (2) branches. One has (from the definition of the coefficient)

$$a = b = e_0/2$$

$$-2e_0 + a + b + e_1 = 0$$

$$-2e_s + e_{s-1} + e_{s+1} = 0 \text{ for } t \geq s \geq 1$$

Thus $e_0 = e_1 = \cdots = e_t = e$.

Next we show $e(x)$ has the prescribed form. We use the notation and results of Chapter 3 of [27].

Let $\beta_3$ have index $r$ and let $s$ be the index of the chain singularity obtained by deleting the first integer on the left from $\beta_3$.

First suppose $\beta_1$ and $\beta_2$ both have index two. Thus

$$\Delta = 2.2.r.l - 2.r - 2.r - 2.2.s = 4(rl - r - s)$$

and so the log discrepancy of $x$ is

$$\frac{2.2.r}{\Delta}(1/2 + 1/2 + 1/r - 1) = 1/(rl - r - s).$$

But the index of $\alpha = rl - s$ and so the spectral value of $\alpha$ is exactly $rl - r - s - 1$.

Thus we may suppose $\beta_1$ has index two, $\beta_2$ has index three, and $3 \leq r \leq 5$. Once again the log discrepancy is

$$\frac{2.3.r}{\Delta}(1/2 + 1/3 + 1/r - 1) = \frac{6-r}{\Delta}.$$

This deals with the case $r = 5$. Otherwise we just need to prove that $\Delta$ is divisible by three in the case $r = 3$ and two in the case $r = 4$. There are two cases, $\beta = (3)$ or $\beta = (2, 2)$.

In the former,

$$\Delta = 2.3.r.l - 5r - 6s = 6rl - 5r - 6s,$$

and in the latter

$$\Delta = 2.3.r.l - 4r - 6s = 6rl - 4r - 6s.$$

The result now follows easily. □

**8.4 Flush divisors in a hunt.**

**8.4.1.** *Let $T$ be a surface. Let $G$ be a reduced curve on $T$, with $K_T + G$ lt. Let $\Gamma$ be a boundary with support $G$. The following implications hold:*

(1) *Assume $K_T + \Gamma$ is level, and $\Gamma$ has a unique component, $E$, whose coefficient, $e$, is minimal. Assume $e < 1$. Let $\Gamma_\epsilon = \Gamma + \epsilon E$. Then $K_T + \Gamma_\epsilon$ is flush for all sufficiently small $\epsilon > 0$.*

(2) *Assume any singularity of $G$ lies along a component of $G$ of minimal coefficient. If $K_T + \Gamma$ is flush at all exceptional divisors of the minimal desingularisation, then $K + t\Gamma$ is flush for any $t > 1$ such that $t\Gamma$ is a pure boundary.*

*Proof.* For (1). Let $e = m(\Gamma)$ be the coefficient of $E$. Let $e' = e + \epsilon$. Of course $e' = m(\Gamma_\epsilon)$. Let $V$ be any exceptional divisor. If $V$ does not lie over $E$ then

$$e(V, K_T + \Gamma_\epsilon) = e(V, K_T + \Gamma) \leq e < e'$$

Since $e' = m(\Gamma_\epsilon)$ it is thus enough to check flushness locally at each point along $E$. At singular points this follows from (8.3.5.2), while at smooth points (of $T$) we can apply (8.3.2.1). Thus (1).

For (2). Let $m = m(\Gamma)$. Note $tm = m(t\Gamma)$. Let $V$ be an exceptional divisor, over the point $x$ in $T$. Suppose first $x$ is a singular point. We can assume $V$ is an exceptional divisor of the minimal desingularisation over $x$, by (8.3.5.4). Let

$$g(t) = e(V, K + t\Gamma) - mt.$$

We have $g(0) \geq 0$, $g(1) < 0$, so $g(t) < 0$ for $t \geq 1$.

Now suppose $x$ is a smooth point. If $G$ is smooth at $x$ then $e(V, K + t\Gamma) < 0$. So assume $x$ is a node of $G$. Then $m$ is the coefficient of one of the components of $G$ through $x$, so we can check flushness locally around $x$, and apply (8.3.2.1). $\square$

**8.4.2 Definition.** For a boundary $\Delta$ on a surface $S$, we let $Z(\Delta)$ be the set of smooth points of $S$ where $\ulcorner\Delta\urcorner$ (the support of $\Delta$) is singular.

**8.4.3 Lemma.** *(Notation as in (8.2.4), (8.2.8)). Suppose $K_S + \Delta$ is anti-ample, and flush away from $Z(\Delta)$. Let $D \subset S$ be the support of $\Delta$.*

*Let $f : T \longrightarrow S$ be the next step of the $K_S + \Delta$ hunt. Assume $\Gamma'$ is a pure boundary. Then*

(1) *$K_T + \Gamma_\epsilon$ and $K_T + \Gamma'$ are flush away from $Z(\Delta)$.*

*If $T$ is not a net, let $\overline{\Sigma} = f(\Sigma) \subset S$. The following hold:*

(2) *If $\overline{\Sigma}$ is a component of $D$, then $K_{S_1} + \Delta_1$ is not level at $q$, $q$ is a smooth point of $S_1$, $D_1$ does not have normal crossings at $q$, and $K_{S_1} + \Delta_1$ is flush away from $Z(\Delta_1)$. Finally, if $K_S + \Delta$ is flush around $\overline{\Sigma}$ then $\Sigma$ is the only exceptional divisor over $q$ at which $K_{S_1} + \Delta_1$ fails to be level.*

(3) *If $\overline{\Sigma}$ is not a component of $D$, and $\overline{\Sigma}$ is disjoint from $Z(\Delta)$ then $K_{S_1} + \Delta_1$ is flush away from $\pi(Z(\Delta))$, and $K_T + \tilde{D} + E$ is log terminal in a neighbourhood of $\Sigma$.*

*Proof.* Note the final remark in (3) follows from previous statements of (3), by (8.0.4) and the definition of $Z(\Delta)$. Note also that the assumptions and conclusions in (1) or (3) are such that we can shrink $S$, and assume $Z(\Delta)$ is empty, that is that $D \cap S^0$ is smooth.

$K_{T_1} + \Gamma$ is level by (8.3.5.4) and (8.3.1.1). By the definition of flush, $e$, the coefficient of $E$ in $\Gamma$, is strictly smaller than the coefficient of any other component of $\Gamma$. So (1) follows from (8.4.1.1) followed by (8.4.1.2) (applied to $\Gamma_\epsilon$).

(3) is immediate from (8.3.1.2).

For (2). Suppose $\overline{\Sigma}$ is a component of $D$. Let $a$, $e'$ be the coefficients of $\Sigma$ and $E$ in $\Gamma'$. Then

$$a = e(\Sigma, K_{S_1} + \Delta_1) > e' = e(\pi(E), K_{S_1} + \Delta_1).$$

Thus $K_{S_1} + \Delta_1$ is not level at $q$. $K_T + \Gamma'$ is flush at any exceptional divisor of the minimal desingularisation of $T$, thus $q$ is not a smooth point by (8.3.5.3). $D_1$ cannot have normal crossings at $q$, by (8.3.2.1). The final remark follows from (8.3.1.2). $\square$

*8.4.4 Remarks.*

(1) When we are hunting for a tiger, we can always assume $\Gamma'$ is a pure boundary by (8.2.5.8).

(2) By definition of the hunt, $f(E)$ is a singular point, while $Z(\Delta)$ is contained in the smooth locus, so we can think of $Z(\Delta)$ as a subset of $T$. Note also that in the conclusion to (3), $q \notin \pi(Z(\Delta))$.

(3) Finally, in (3) the condition $\overline{\Sigma} \cap Z(\Delta) = \emptyset$ is satisfied whenever the multiplicity of $\Delta$ (that is the sum of the coefficients, weighted by the multiplicities of the components see (21.2)) is at least one at each point of $Z(\Delta)$. This will always be the case in any hunt we consider.

**8.4.5 Lemma.** *In the hunt with $\Delta_0 = \emptyset$, $K_{T_1} + eE_1$ is flush. If $T$ is not a net, then $K_{S_1} + a_1A_1$ is flush.*

*Proof.* Follow the proof of (1) and (3) of (8.4.3). □

**8.4.6 Corollary.** *In the hunt with $\Delta_0 = \emptyset$, assume that $Z(\Delta_i) \cap \overline{\Sigma_{i+1}} = \emptyset$ for all $i > 0$ (see (8.4.4.3)). Then $K_{S_i} + \Delta_i$ is flush away from $Z(\Delta_i)$ for all $i > 0$.*

*Proof.* Immediate from (8.4.3) and (8.4.5), by induction. □

**8.4.7 Proposition: General configuration of the first two steps of the hunt with $\Delta_0 = \emptyset$.** *Notation as in (8.2.10).*

*For the first hunt step:*

*$K_{T_1} + E_1$ is log terminal, $K_{T_1} + \Gamma'$ is flush and one of the following holds:*

(1) *$T_1$ is a net.*

*otherwise $K_{S_1} + a_1A_1$ is flush and one of the following holds*

(2) *$g(A_1) > 1$ (arithmetic genus).*
(3) *$g(A_1) = 1$ and $A_1$ has an ordinary node at $q = q_1$.*
(4) *$g(A_1) = 1$ and $A_1$ has an ordinary cusp at $q = q_1$.*
(5) *$g(A_1) = 0$ and $K_{S_1} + A_1$ is log terminal.*

*For the second hunt step one of the following holds:*

(6) *$T_2$ is a net*
(7) *$A$ is contracted by $\pi_2$ (that is $A = \Sigma_2$), $K_{T_2} + \Gamma'$ is flush, $K_{S_1} + A_1$ is log terminal, $q_2$ is a smooth point of $S_2$, $B_2$ is singular at $q_2$, with a unibranch singularity, and $K_{S_2} + \Delta_2$ is flush away from $q_2$, but is not level at $q_2$. $\Sigma_2$ is the only exceptional divisor at which $K_{S_2} + \Delta_2$ fails to be flush.*
(8) ***$\Delta_2$ has two components.***

*Suppose in (8) that $a_2 + b_2 > 1$. Then:*

*(8.1): $\overline{\Sigma}_2 \cap \mathrm{Sing}(A_1) = \emptyset$, $K_{T_2} + \Gamma'$ is flush away $\mathrm{Sing}(A_1)$. $K_{S_2} + \Delta_2$ is flush away from $\pi_2(\mathrm{Sing}(A_1))$, and at least one of $-(K_{S_2} + A)$ or $-(K_{S_2} + B)$ is ample. Also, one of the following holds:*

(9) *$(S_2, A + B)$ is a fence.*
(10) *$(S_2, A + B)$ is a banana, $K_{S_2} + B$ is plt, and $x_1 \in A$.*

(11) $(S_2, A+B)$ *is a tacnode, with tacnode at* $q_2$. $K_{S_2}+B$ *is plt. If* $x_1 \in A$, $A \cap B = \{x_1, q_2\}$. *If* $x_1 \notin A$ *then* $A \cap B = \{q_2\}$.

*Remark.* In (8.1), $\mathrm{Sing}(A_1)$ (which is either empty, or $q_1$) is a smooth point of $S_1$ and so $f_1$ is an isomorphism around $\mathrm{Sing}(A_1)$, thus we can think of $\mathrm{Sing}(A_1)$ as a subset of $T_2$.

*Proof of (8.0.14).* Everything through (5) is clear, using (8.4.3).

Consider the second hunt step, and suppose $T_2$ is not a net.

If $A$ is contracted, then obviously $A = \Sigma_2$, and we have case (5) of the first hunt step. The rest follows from (8.4.3.2).

Now suppose $\Delta_2$ has two components, and $a_2 + b_2 \geq 1$. In particular $a_2 > 1/2$, and so $\Sigma_2$ cannot meet a singular point of $A$. All the flush remarks follow from (8.4.3.3). $A_2 + B_2$ cannot have a triple point at $q_2$, by (8.3.7). So $A \cap E_1 \cap \Sigma_2 = \emptyset$, and $A \cap \Sigma_2$, $E_1 \cap \Sigma_2$ are each at most one point (and the second cannot be empty as $B^2 > 0$). By (8.3.5.1), $K_{S_2} + A + B$ is lt at singular points of $S_2$. If $X$ is the smaller of $A$ and $B$, then $-(K + (a_2 + b_2)X)$ is ample. The rest follows from easy set theoretic considerations. □

## §9 Bogomolov Bound

This section is logically independent from the rest of the paper and is (we hope) of independent interest.

We refer to Chapter 10 of [27] for the definition and basic properties of $\mathcal{Q}$-sheaves and $\hat{\Omega}^1_S$.

The main result of this section is the following:

**9.1 Theorem.** *Suppose* $B \subset S$ *is a reduced curve on a surface of Picard number one, and* $K_S + B$ *is lc. Then* $\hat{c}_2(\hat{\Omega}^1_S(\log B)) \geq 0$, *with strict inequality when* $K_S + B$ *is ample.*

*Proof.* By (10.14) of [27] we can assume $K_S + B$ is anti-ample. We slightly modify an argument due to Kawamata [20]. We may assume $\hat{\Omega}^1_S(\log B)$ is not semi-stable. Thus there is a saturated destabilising $\mathcal{Q}$ subsheaf $\mathcal{L}$, cf. the proof of (10.11) of [27]. Suppose $c_1(\mathcal{L}) = t(K_S + B)$. The crucial observation is that $h^0(S, n\mathcal{L}) < cn$ for some constant $c$, cf. the proof of (10.14) of [27]. Thus $\mathcal{L}$ is not ample, and $t \leq 0$. Now proceed, exactly as in [20]. □

**9.2 Corollary.** *Let* $S$ *be a log terminal surface of Picard number one and* $B \subset S$ *a reduced curve such that* $K_S + B$ *is lc. Let* $U = S \setminus B$.

$$\sum_{p \in U} \frac{r_p - 1}{r_p} \leq e_{\mathrm{top}}(S) - e_{\mathrm{top}}(B)$$

*If $S$ is rational, and $B$ has arithmetic genus 0 and $s$ irreducible components, then*

$$\sum_{p \in U} \frac{r_p - 1}{r_p} \begin{cases} \leq 3 & \text{if } s = 0 \\ \leq 1 & \text{if } s = 1 \\ = 0 & \text{if } s = 2. \end{cases}$$

*Proof.* The first inequality follows from (10.7) and (10.8) of [27]. If $S$ is rational, then $e_{\text{top}}(S) = 3$. $B$ is a connected tree of $\mathbb{P}^1$s. If $B$ is not empty, then $e_{\text{top}}(B) = 1 + s$. Hence the second set of inequalities follows from the first inequality. □

**9.3 Corollary.** *Fix $\epsilon > 0$, and a birational equivalence class $\mathcal{C}$ of projective surfaces. The collection*

$$X = \{\, S \in \mathcal{C} \,|\, \rho(S) = 1, e(S) < 1 - \epsilon \,\}$$

*is bounded.*

*Proof.* Note in particular that $e(S) < 1$, so $S$ has quotient and thus rational singularities. For $S \in \mathcal{C}$ with rational singularities, $e_{\text{top}}(S)$ depends only on $\rho(S)$. Thus (9.2) implies any surface $S \in X$ has at most $2e_{\text{top}}(S)$ singularities. By the classification of quotient singularities, it is enough to bound $\rho(\tilde{S})$ for $S \in X$, and thus to bound $\{\, -K_{\tilde{S}}^2 \,|\, S \in X \,\}$ from above, where $f : \tilde{S} \longrightarrow S$ is the minimal desingularisation.

Write $f^*K_S = K_{\tilde{S}} + \sum e_i E_i$, and let $r_i = -E_i^2$. We have

$$\begin{aligned} -K_{\tilde{S}}^2 &= -K_S^2 + \sum e_i K_{\tilde{S}} \cdot E_i \\ &\leq \sum e_i (r_i - 2) \end{aligned}$$

Since $e_i < 1$, it is enough to bound the number of $E_i$ with $r_i \geq 3$. This is bounded in terms of $\epsilon$ by (3.1.12) of [27] and simple coefficient calculations. □

**An interesting example in non-zero characteristic.**

Note in particular, the Bogomolov bound implies a rank one log Del Pezzo can have at most five singular points (there is no rank one log del Pezzo with exactly six $A_1$ singularities). Here we show there is no bound on the number of singularities in characteristic two.

Take a strange conic in $\mathbb{P}^2$, and blow up the point where all the tangents meet. Now we have a curve $C$ in $\mathbb{F}_1$, which is purely inseparable over the base of degree two.

Pick any fibre and perform the following operation. Blow up along $C$ twice and contract the two $-2$-curves. If we perform this operation $k$ times, $C$ will now have self-intersection $4 - 2k$, and will lie in the smooth locus. Contracting $C$ we get a log del Pezzo of rank one, with $2k + 1$ singular points. As a special case ($k = 3$) we get a Gorenstein log del Pezzo with seven nodes.

## §10 Riemann Roch and Surfaces with Small Coefficient

We are seeking to partially classify those rank one log del Pezzo surfaces with coefficient less than 2/3 which do not have tigers. Our method is simply to use Riemann Roch and the Bogomolov bound. In fact for our classification will require much less than the absence of a tiger, see (10.8).

It will be convenient (at various points in the paper) to have the following list of singularities of small coefficient:

**10.1 Proposition.** *Let $x$ be an lt singularity, with $e = e(x)$. If $0 < e < 3/5$, the possibilities for $x$ are as follows:*

*(1)* $e < 1/2$: $(3, A_j)$. $e = \frac{j+1}{2j+3}$.

*(2)* $e = 1/2$:

*(a)* $(4)$

*(b)* $(3, A_j, 3)$

*(c)* $(2, 3, 2)$

*(d) $x$ is a non-chain singularity, with center $(2)$ and branches $(2)$, $(2)$, and $(A_j, 3)$, with the central curve and the unique $-3$-curve meeting opposite ends of the $A_j$ chain. This is the only non-chain singularity with $e < 2/3$.*

*(3)* $1/2 < e < 3/5$:

*(a)* $(2, 3, A_j)$ *with* $2 \le j \le 4$. $e = \frac{2j+2}{3j+5}$.

*(b)* $(4, 2)$, $e = 4/7$.

*Proof.* (L.1) reduces the result to straightforward checking. □

**10.2 The following notation is fixed throughout the section.**

*$S$ is a rank one log del Pezzo, and $\tilde{S} \longrightarrow S$ is the minimal desingularisation. $\rho$ is the relative Picard number of $\tilde{S} \longrightarrow S$.*

*For a singular point $p \in S$, $\rho(p)$ indicates the contribution of $p$ to $\rho$, that is the number of exceptional divisors lying over $p$.*

*Let $r : \overline{S} \longrightarrow S$ be the minimal resolution of $S$ over the non Du Val locus. Thus $r$ resolves precisely the non Du Val singularities. In particular, $\overline{S}$ has Gorenstein singularities.*

*Let $D \subset \overline{S}$ be the (reduced) sum of the $r$-exceptional divisors, and $n$ the number of non Du Val singularities. Let $\Gamma = K_{\overline{S}} \cdot \sum(1 - e_i)E_i$, where the sum is over the $r$-exceptional divisors, and $e_i = e(E_i, K_S)$.*

*Since $H^0(2K_{\overline{S}} + D) = 0$ we have by Riemann Roch*

$$\begin{aligned} h^0(-K_{\overline{S}} - D) &\geq \chi(-K_{\overline{S}} - D) = 1 - n + K_{\overline{S}}^2 + \Gamma \\ &= 1 - n + K_{\overline{S}}^2 + K_{\overline{S}} \cdot D \end{aligned} \tag{10.3}$$

*For a (non Du Val) singular point t we let $\Gamma(t)$, $\delta(t)$, and $w(t)$ be the contribution of t to $\Gamma$, $-K_{\bar{S}}^2$, and $K_{\overline{S}} \cdot D$ respectively. Thus*

$$\Gamma(t) = \sum(1 - e_j)K_{\tilde{S}} \cdot E_j = \sum(1 - e_j)(-2 - E_j^2)$$
$$\delta(t) = \sum e_j(-2 - E_j^2)$$
$$w(t) = \sum K_{\tilde{S}} \cdot E_j = \sum(-2 - E_j^2)$$

*where in each case the sum is over those j with $E_j$ lying over t. We also refer to $w(t)$ as the* **weight** *of t. Let w be the sum of the weights of the singular points, and $\beta$ be the weighted average of the coefficients, that is*

$$w \cdot \beta = \sum w(t)e(t).$$

*Observe*

$$\beta \leq (1/w)\sum w(t)e(S) = e(S).$$

*We let $\Delta(t)$ be the index of t (that is the order of the local fundamental group).*

*In the hunt with $S = S_0$, we let $\alpha$ be sum of $\delta(t)$ for $t \notin \overline{\Sigma}$. Recall from (8.2.10) that $\overline{\Sigma} \subset S_0$ is the image of $\Sigma_1 \subset T_1$.*

**10.4 Lemma.** *(notation as in (10.2)) If $n \leq 1$, or $n = 2$ and $e(S) \leq 1/2$, then $|-K_{\overline{S}} - D|$ is non-empty, in particular any exceptional divisor over a non Du Val point is a tiger.*

*Proof.* By (10.1), $\Gamma(t) \geq 1/2$, if $e \leq 1/2$. Since $K_{\bar{S}}^2 > 0$, $|-K_{\overline{S}} - D|$ is non-empty, by (10.3). □

It is convenient to list all four-tuples of integers at least 3, which can occur as indices on a rank one log del Pezzo (that is integers satisfying the Bogomolov bound). We note in each case that by the Bogomolov bound there can be no other singularities on the surface.

**10.5**

(a) $(3, 3, 3, m)$, $m \geq 3$, or
(b) $(3, 3, 4, m)$, $4 \leq m \leq 12$, or
(c) $(3, 3, 5, m)$, $5 \leq m \leq 7$, or
(d) $(3, 3, 6, 6)$, or
(e) $(3, 4, 4, m)$ $4 \leq m \leq 6$, or
(f) $(4, 4, 4, 4)$.

**10.6 Lemma.** *Notation as in (10.2). Let S be a rank one log del Pezzo such that $|-K_{\overline{S}} - D|$ is empty. Suppose S has m singularities of index at least three.*

*If $m > 3$, then the singularities of S are, either*

(1) *$(3)$, $(3)$, $(2, 2, 2)$ and $(3, 2, 2, 2, 2)$, or*

(2) (3), (3), (3, 2) *and an* $A_6$*-point, or*

(3) *the indices are given by (10.5.a).*

*Proof.* Suppose not.

Note first that by the classification of quotient singularities [7] there is only one non cyclic singularity, $D_4$, of index (that is order of local fundamental group) less than 16.

Using (L.1) it is simple to classify singularities of small index. For example the only non Du Val singularities of index at most 7 are

(10.7) $$(7), (6), (5), (4), (3), (4,2), (3,2), (3,2,2).$$

We will use this, as well as (10.1), repeatedly and without remark.

Clearly $m = 4$, using the Bogomolov bound. We consider the above list of possible indices. Let $z$ be the fourth singularity in the list. We first show that cases (d)-(f) above do not occur. Note $\rho \geq 10$, for otherwise $K_{\bar{S}}^2 \geq 0$ and $|-K_{\overline{S}} - D| \neq \emptyset$ by (10.3).

In case (d), one of the index six points must be $A_5$ (or $\rho \leq 6$) the other type (6) (otherwise $n \leq 2$ and $e_0 \leq 1/3$ violating (10.4)). But then both index three points are Du Val (or $\rho \leq 9$) and $n = 1$, contradicting (10.4).

In case (e), the points other than $z$ cannot all be Du Val (or $n \leq 1$ violating (10.4)). Thus $\rho(z) \geq 3$. Thus $z$ is Du Val. Since $n \geq 2$, $\rho \leq 9$ unless the singularities are (3),(4), $A_3$ and $z = A_5$. This violates (10.4).

In case (f), $n \geq 3$ by (10.4), thus $\rho \leq 6$, a contradiction.

In case (b), consider $w(z)$. If it is zero, then one checks $(n-1) - \Gamma \leq 0$, contradicting (10.3). So $w \geq 1$. $n > 1$ by (10.4). So $\rho - \rho(z)$ is at most 6. Thus $\rho(z) \geq 4$. From $\Delta(z) \leq 12$ it follows that $z = (3, A_j)$. Thus by (10.4), $n \geq 3$. (1) is the only possibility with $\rho \geq 10$.

In case (c), suppose first $z$ is non Du Val. Since $n \geq 2$, $\rho - \rho(z) \leq 7$, thus $\rho(z) \geq 3$, which implies $z = (3,2,2)$. It follows the singularities are (3), $A_2$, $A_4$ and $(3,2,2)$. This violates (10.4). Thus $z$ is Du Val. Then one checks $(n-1) - \Gamma \leq 0$ unless we have (2). □

**10.8 Lemma.** *Notation as in (10.2). Assume* $e(S) < 2/3$ *and* $|-K_{\overline{S}} - D| = \emptyset$. *We have the following possibilities. Either*

(1) $n = 4$, $w \geq 4$, *three of the singularities are* (3), *or*

(2) $K_{\bar{S}}^2 = -3$, $n = 3$, $w = 5$, $3/5 < \beta \leq e(S)$, *or*

(3) $K_{\bar{S}}^2 = -2$, $n = 3$, $w = 4$, $1/2 < \beta \leq e(S)$, *or*

(4) $K_{\bar{S}}^2 = -1$, $n = 3$, $w = 3$, $1/3 < \beta \leq e(S)$, *or*

(5) $K_{\bar{S}}^2 = -1$, $n = 2$, $w = 2$, $1/2 < \beta \leq e(S)$.

*Furthermore, if* $e(S) < 1/2$ *then* $K_{\bar{S}}^2 = -1$, $e > 1/3$, *and either (4) holds, or (1) holds, and the last singularity is* $(3, A_6)$.

*Proof.* By the Bogomolov bound, $n \le 4$. By (10.3) and the definitions we have the following formulae:

$$0 < -K_{\tilde{S}}^2 < 2w/3, \quad w \ge n$$

$$w \cdot \beta \ge \sum e_i K_{\overline{S}} \cdot E_i = K_{\tilde{S}}^2 - K_{\overline{S}}^2 > -K_{\overline{S}}^2$$

$$\Gamma > w/3, \quad \beta \le e, \quad \Gamma < n-1, \quad -K_{\tilde{S}}^2 \ge w - n + 1.$$

If $n = 4$ we have (1) by (10.6).

Suppose $n = 2$. Then $w/3 < \Gamma < 1$. Thus $w = 2$ and $K_{\tilde{S}}^2 = -1$.

Suppose $n = 3$. $w/3 < \Gamma < 2$ so $w < 6$. The corresponding values of $K_{\tilde{S}}^2$, and $\beta$ follow from the formulae above.

Finally suppose $e(S) < 1/2$. Then by (10.1), $w = n \le 4$, and $0 < -K_{\tilde{S}}^2 < w/2 \le 2$. Thus $K_{\tilde{S}}^2 = -1$, and $\beta \le e < 1/2$. So if we do not have (4), then $n = w = 4$. The last singularity has $\rho = 7$, and thus by (10.1) is $(3, A_6)$. □

Here we house some simple results, which might otherwise be on the streets.

**10.9 Lemma.** *Notation as in (8.2.10).*

$$K_{S_0}^2 = \frac{(K_{S_0} \cdot \Sigma)^2}{\Sigma^2}$$

*Proof.* Immediate since $S_0$ has rank one. □

**10.10 Lemma.** *Let $(S, p)$ be a lt germ, and $C \subset S$ an analytically irreducible curve germ at $p$. Suppose $p$ has weight one, and $\tilde{C}$ meets only one exceptional divisor, $E$, of the minimal desingularisation over $p$, the unique $-3$-curve over $p$, and meets $E$ normally. Then the coefficient of $E$ in the pullback of $C$ is $e(p)$.*

*Proof.* Follows since $\tilde{C}$ and $-K_{\tilde{S}}$ have the same intersection with each exceptional divisor of the minimal desingularisation over $p$. □

**10.11 Lemma.** *Let $x \in S$ be a cyclic non Du Val singularity. Then there is an exceptional divisor $E \subset \tilde{S}$ of the minimal desingularisation over $x$ of maximal coefficient (for $K_S$) such that $E^2 \le -3$.*

*Proof.* An easy coefficient calculation. □

## §11 A partial Classification of $K_T$-contractions

The goal of this section is to give a partial classification for the contractions $\pi_i$ in the hunt. We first consider birational contractions, (11.0-4), and then $\mathbb{P}^1$-fibrations in (11.5).

**11.0 Birational Contractions.**

Our main objective is a local analytic classification of the possibilities for

$$\pi_i : (T_i, \Sigma_i + D_i) \longrightarrow (S_{i+1}, D_{i+1})$$

for the first two hunt steps, $i = 1, 2$. By local analytic, we mean locally analytically about $q_{i+1} \in S_{i+1}$.

As such a classification will be useful at other points in the paper, we will work in the abstract (outside the context of the hunt).

*11.0.1 We fix some notation. Let $\pi : T \longrightarrow S$ be a proper birational contraction of a $K_T$-extremal ray. In particular $\rho(T/S) = 1$. Let $\Sigma$ denote the exceptional divisor of $\pi$, which is a $-1$-curve. Let $q$ be the image of $\Sigma$. We will only be concerned with the local analytic description of $T$ about $\Sigma$, thus $p \in S$ is only an analytic surface germ. Let $W \subset T$ be a curve with smooth components crossing normally. We assume $W$ has at most two irreducible components $X, Y$ ($Y$ is allowed to be empty). We assume $K_T + cX + dY$ is $\pi$-trivial and flush, with $0 < c, d < 1$. Thus by (8.0.4), $K_T + X + Y$ is log terminal. In particular $X$ and $Y$ cannot meet at any singular point of $T$.*

*We assume $\pi|_W$ is finite (that is $\Sigma$ is not a component of $W$). Thus by (8.3.1.2), $K_S + cX + dY$ is flush. We assume $X$ and $Y$ have analytically irreducible images at $q$, and thus each (is either empty or) meets $\Sigma$ at exactly one point.*

*Let $D \subset S$ be the image (with reduced structure) of $W$.*

*At a chain singularity $(a_1, \ldots, a_p)$ a prime (resp. underline) indicates normal contact with $\Sigma$ (resp. $W$). Note this is the same notation we use in the hunt, see (8.2.10).*

*We let $h : \tilde{T} \longrightarrow \tilde{S}$ be the induced map between the minimal desingularisations, a composition of smooth blow ups.*

We will also use the symbols $X$, $Y$ to indicate the images of these curves on $S$. This is in keeping with our convention of occasionally using the same symbol to denote a divisor, and its strict transform under a birational map.

The starting point for our classification is the following simple observation:

**11.0.2 Lemma.** *The centre of each blow up of $h$ lies on the exceptional divisor of the previous blow up.*

*Proof.* $\tilde{\Sigma}$ is the only $-1$-curve lying over $q$. □

We classify configurations by considering possibilities for the sequence of blow ups $h$. We think of $(T, W)$ as being obtained from $(S, D)$ by the sequence of blow ups $h$.

We note next that (under some mild conditions) the coefficients $c$, $d$ do not decrease with additional blow ups:

**11.0.3 Lemma.** *If $(T', X', Y', \Sigma')$ is obtained from $(T, X, Y, \Sigma)$ by further blow ups, that is if $h' : \tilde{T}' \longrightarrow \tilde{S}$ factors through $h : \tilde{T} \longrightarrow \tilde{S}$ then*

$$(c'X + d'Y) \cdot \Sigma \geq (cX + dY) \cdot \Sigma$$

*In particular if $c = d$ and $c' = d'$, or $d' \leq d$, then $c' \geq c$.*

*Proof.* Let $f$ be the induced map $\tilde{T}' \longrightarrow \tilde{T}$. Let $p$, $p'$ be the minimal desingularisation maps. Note that

$$p_* f_* p'^* (K_{T'} + c'X' + d'Y') = K_T + c'X + d'Y + \alpha\Sigma$$

for some $\alpha \geq 0$, and that the above divisor is pulled back from $S$, and thus is $\Sigma$ trivial. Hence

$$\begin{aligned}(c'X + d'Y) \cdot \Sigma &= -\alpha\Sigma^2 - K_T \cdot \Sigma \\ &= -\alpha\Sigma^2 + (cX + dY) \cdot \Sigma.\end{aligned}$$

As $\Sigma$ has negative self-intersection, the result follows. □

When $D$ is smooth at $q$, then $K_S + D$ is log terminal at $q$, by (8.0.4). We will consider this case in (11.4). Suppose $D$ is singular at $q$. By (8.0.4), $S$ is smooth. Let $g$ be the local arithmetic genus of $D$. We first classify those $D$ with a double point, (11.1-2). Higher multiplicity is considered in (11.3). Double points for curves on a smooth surface germ are classified in II.8 of [4]. There are two possibilities, and we consider each in turn.

**11.1** $D$ **has two smooth branches at** $q$**.**

In this case, $D$ has what is called an $A_{2g}$-singularity. $X \subset S$ and $Y \subset S$ are both smooth, and meet to order $g$ at $q$. We call this a **node of order** $g$**.** Consider $h : \tilde{T} \longrightarrow S$. If $g \geq 2$, we must first blow up repeatedly at $X \cap Y$ until $X$ and $Y$ cross normally, for otherwise $K + W$ will not be lt. We call this **configuration** 0. Here $X$ and $Y$ meet normally at a smooth point of $T$ along $\Sigma$. $\Sigma$ contains a unique singular point, $(2', \ldots, 2) = A_{g-2}$ (and $g \geq 2$). Clearly $c + d = 1$.

*Remark.* Note that in this case $W$ is singular, so configuration 0 cannot occur for $T = T_1$ of the hunt, with $W = E_1$, since $E_1$ is smooth.

If there is a further blow up it must be at $X \cap Y$, or otherwise $X$ and $Y$ will meet at a singular point of $T$. We call this **configuration I.** Here $X$ and $Y$ each meet $\Sigma$ at a smooth point, there is one singular point $(2', 2, \ldots, 2) = A_{g-1}$ and once again $c + d = 1$.

Consider the next blow up. It must be at the point where $\Sigma$ meets either $X$ or $Y$, for if we blow up at any other point, then $X$ and $Y$ will meet at a singular point of $T$. Switching $X$ and $Y$ if necessary, we may assume the next blow up is at the point of $X$.

We call this **configuration** ***II.*** Here there is a unique singularity, $A_g$ along $\Sigma$, $X$ meets $\Sigma$ at a smooth point. On $\tilde{T}$, $Y$ and $\Sigma$ are disjoint, but meet the same $-2$-curve of the $A_g$ chain, an end of the chain. The $A_g$ point has type $(\underline{2'}, \ldots, 2)$. One computes $gd + (g+1)c = g + 1$.

Now at every further step there are two possible points we could blow up, either the point $x \in \Sigma \subset \tilde{T}$ lying over $\Sigma \cap X \in T$, or the point $y \in \Sigma \subset \tilde{T}$ lying over $\Sigma \cap Y \in T$ (if we blow up at any other point, then $X$ and $Y$ will meet at a singular point of $T$). We call these the points **nearest** $X$ and $Y$ respectively. We indicate the sequence of blow ups by a string of $x$s and $y$s. For example $(II, x, x, y)$ means the configuration obtained from $II$ by blowing up twice at $x$ and once at $y$.

Suppose $g \geq 2$. Note that if any of the blow ups is at the point nearest $X$, then $K_T + Y$ is not log terminal. Thus for $g \geq 2$ the only possibilities are $(II, y, \ldots, y)$, which we indicate by $(II, y^s)$, where $s$ is the number of $y$s in the string.

Figure 5 gives a picture of some possible configurations:

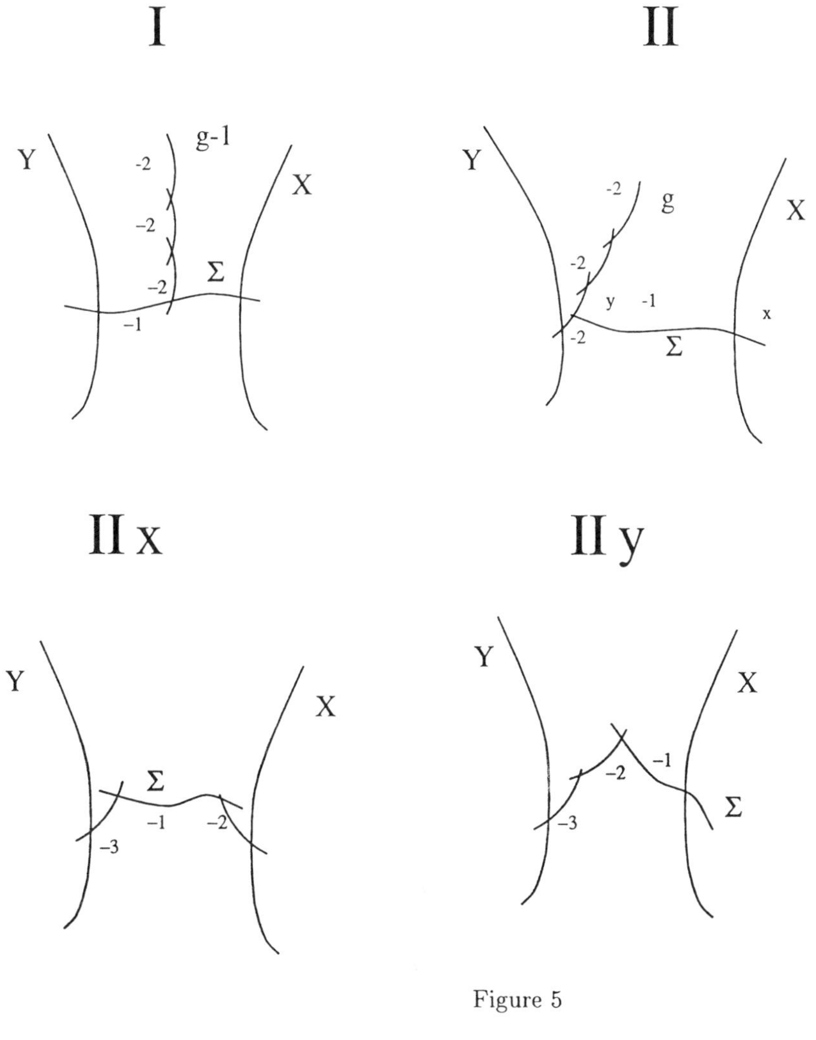


Figure 5

*11.1.0 Remark.* The above classification of configurations (ignoring the values $c$ and $d$) requires only the assumption that $K_T+X+Y$ is lt (that is it holds without assuming either $K+cX+dY$ is flush, or $\pi$-trivial), as this is all we have used to this point. Using the additional assumptions we can further restrict the possibilities:

**11.1.1 Lemma.** *Suppose $D$ has a node of order $g$.*

*One of the following holds:*

(1) *$T$ has type I or 0, and $c+d=1$, or*

(2) *$g \geq 2$, $T$ has type II, and $gd+(g+1)c=(g+1)$, or*

*$g=1$, $K_T+\Sigma+X+Y$ is log canonical and either*

(3) *$T$ has type $(II, x^{r-1})$, with $r \geq 1$ (for $r=1$ this means configuration II) there is a unique singularity, an $A_r$ point, on $\Sigma$, $X$ meets $\Sigma$ at a smooth point, $Y$ meets $\Sigma$ at the $A_r$ point, which has type $(\underline{2},\ldots,2')$, and $c+\frac{d}{r+1}=1$, or*

(4) *$\Sigma$ meets $X$ and $Y$ each at a singular point of $T$, and those are the only two singularities along $\Sigma$.*

*Moreover*

(5)

$$g(c+d-1) < \min(c,d) < \frac{g}{2g-1}.$$

*Proof.* Let $E$ be the $-1$-curve of configuration I. One computes $e(E, K+cX+dY) = g(c+d-1)$. (5) then follows from the flush condition.

The equalities in the lemma are an easy computation.

Suppose $g \geq 2$. In configurations $(II, y^s)$, $Y$ has spectral value at least two, and so $c, d \geq 2/3$, by (8.0.7.2), which contradicts (5).

The rest is easy. □

**11.2 $D$ has a unibranch double point at $q$.**

We describe this as a $g^{th}$ **order cusp.** In this case $Y$ is empty and $W=X$. We again consider $h$. The first series of blow ups must remove the singularity of $D$ (since $K_T+X$ is log terminal). We call this **configuration I.** Here $\Sigma$ has one singularity, $(2',\ldots,2)=A_{g-1}$, and $X$ is simply tangent to $\Sigma$ at a smooth point.

The next blow up must be at $X \cap \Sigma$. Call this **configuration II.** Here there is an $A_g$ singularity of type $(\underline{2}',2,\ldots 2)$.

The next blow up is again along $X$. Call this **configuration III.** Then $\Sigma$ has two singularities $(A_{g-1}, 3')$ and an $A_1$ point, and $X$ meets $\Sigma$ at a smooth point. Let $u$, $v$, $w$ be the points where

$\Sigma$ meets the $-3$-curve, $X$, and the $-2$-curve. The next blow up is at one of these points, and we call the resulting configurations $u$, $v$ and $w$.

In $u$ there are two singular point along $\Sigma$, $(A_{g-1}, 4')$ and $(\underline{2'}, 2)$.

In $v$ there is a single singular point $(A_{g-1}, 3, 2', 2)$ along $\Sigma$, and $\Sigma \cap X$ is a smooth point of $T$.

In $w$ there are two singular point along $\Sigma$, $(A_{g-1}, 3, \underline{2'})$ and $(3')$.

Beyond these there are two possible points to blow up. We indicate by a string of $ns$ and $fs$ whether we blow up at the point nearest, or farthest from $X$ (that is at a point of $\tilde{T}$ lying over $X \cap \Sigma \subset T$, or a point of $\Sigma$ not lying over $X \subset T$).

Figure 6 gives a picture of some possible contractions:

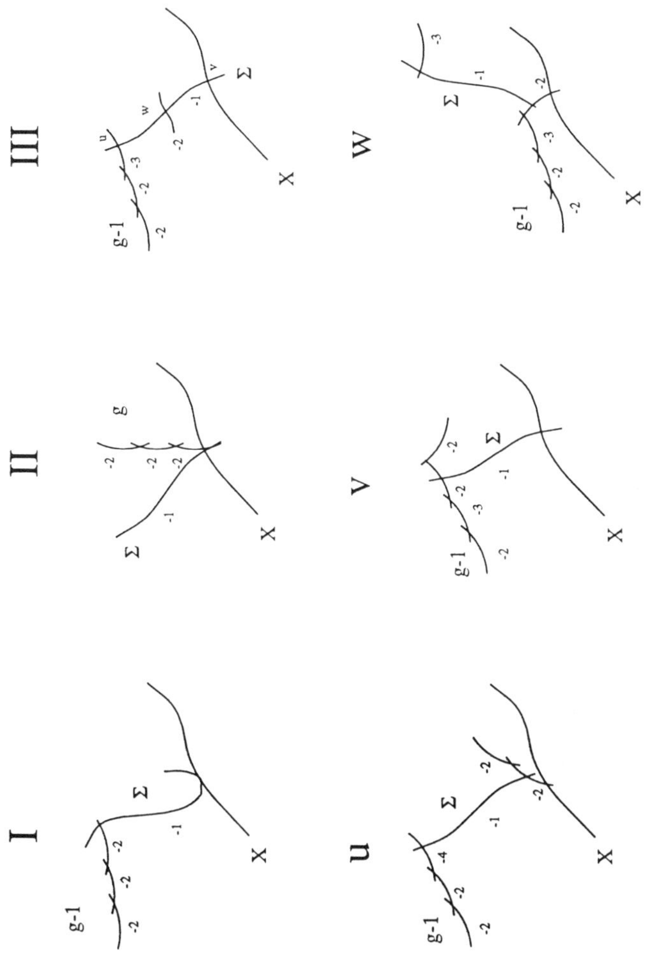

Figure 6

*11.2.0 Remark.* The above classification requires only the assumption that $K_T + X$ is lt (that is it holds without assuming either $K + cX$ is flush, or $\pi$-trivial), as this is all we have used to this point. Using the additional assumptions we can further restrict the possibilities:

**11.2.1 Lemma.** *Suppose $D$ has a cusp of order $g$. Then, either*

(1) *$T$ has type I, and $c = 1/2$, or*
(2) *$T$ has type II, and $c = (g+1)/(2g+1)$, or*
(3) *$T$ has type III, and $c = (g+1)/(2g+1)$, or*
(4) *$g = 1$, $T$ has type $u$ and $c = 3/4$ or $g = 2$ and $c = 9/14$, or*
(5) *$g = 1$, $T$ has type $v$ and $c = 5/7$ or $g = 2$ and $c = 7/11$, or*
(6) *$g = 1$, $T$ has type $w$ and $c = 7/9$, or*
(7) *$g = 1$, $T$ has type $(u; n)$ and $c = 11/14$, or*
(8) *$g = 1$, $T$ has type $(v; f)$ and $c = 10/13$, or*
(9) *$g = 1$, $T$ has type $(v; f^2)$ and $c = 15/19$, or*
(10) *$g = 1$, $T$ has type $(v; n)$ and $c = 3/4$, or*
(11) *$g = 1$, $T$ has type $(v; n^2)$ and $c = 7/9$.*

*Moreover*

(12) $c \leq \frac{g}{2g-1}$
(13) *If $g = 1$, $c \leq 4/5$.*

*Proof.* The stated values of $c$ are elementary calculations. Thus if we have type I, II or III, then we have the equalities of (1)-(3). Let $E$ be the $-1$-curve of configuration I. One computes $e(E, K + cX) = g(2c - 1)$, so (12) follows from the flush condition. (13) follows by computing the coefficient of the $-1$-curve of configuration *III*.

Suppose we have type $u$. If $g \geq 3$, then we have a singularity at least $(4, 2, 2)$ which has coefficient $3/5$, which contradicts (12). Further blow ups are of form $(u; n^r)$ (or $K + X$ is not lt). If $g \geq 2$, then we have a singularity at least $(2, 4, 2)$ which has coefficient $2/3$. If $g = 1$ and $r \geq 2$ we get a singularity of spectral value four, impossible, by (13), and (8.0.7.2).

Suppose we have type $w$. If $g \geq 2$, then $E$ has a singularity of spectral value at least 2, impossible by (12). Further blow ups are of the form $(w; n^r)$. Then $X$ has a singularity of spectral value at least four, violating (13).

Suppose we have type $v$. If $g \geq 3$ then we have a singularity at least $(2, 2, 3, 2, 2)$ which has coefficient $3/5$, contradicting (12). For $(v; n)$ or $(v; f)$ or beyond, if $g \geq 2$ then there is a point of coefficient at least $2/3$, contradicting (12). So $g = 1$. $c(v; f) = 10/13$. $c(v; f^2) = 15/19$. In $(v; f, n)$, $(v; f^2, n)$, $(v; f^3)$, $(v; n, f)$, $(v; n^2, f)$, or $v(n^3)$, $f$ has coefficient at least $4/5$, so these and further blow ups are not allowed. There are no remaining cases to consider. □

**11.3 $D$ has multiplicity three.** Our only application will be to the first hunt step when $A$ has multiplicity three. (We do not need to consider higher multiplicities in the hunt, by (8.0.10)). A priori $A$ might have three analytic branches at $q$. Hence *We broaden the set up of (11.0.1) and allow $W$ to have a third component, $Z$, and consider the log divisor $K_T + cX + dY + fZ$, which we assume is $\pi$-trivial and flush as before.* In fact we will show shortly that $Z$ is empty.

Let $e$ be the smallest non-zero coefficient in $cX + dY + fZ$. $e < 1/2$ by (8.3.7). Thus by (8.0.7.2) and (8.0.8), $W$ is contained in the Du Val locus of $T$.

**11.3.0 Lemma.** *Hunt notation as in (8.2.10). Suppose $S_0$ does not have a tiger, and $A \subset S_1$ has multiplicity at least three. Then $1/3 < a < 1/2$. Furthermore, if $E_1 \subset T_1$ contains an $A_k$ point, then $x_0 = (3, A_k)$, $E_1$ is the $-3$-curve, and $e_0 = (k+1)/(2k+3)$.*

*Proof.* $S_0$ is not Du Val, for example by (8.0.9). $e_0 < a < 1/2$ by (8.3.7). Now the result follows from the list of singularities of small coefficient, (10.1). □

In view of (11.3.0) we make the additional assumption:

*To simplify the classification we will assume that $e > 1/3$ and furthermore if $W$ contains an $A_k$ point, then $e > (k+1)/(2k+3)$.*

We again study the composition of blow ups $h : \tilde{T} \longrightarrow \tilde{S}$. We write $\Sigma^i$ to indicate the $-1$-curve of the $i^{th}$ blow up.

**11.3.1 Lemma.** *$Z$ is empty, that is $D$ has at most two branches.*

*Proof.* As we have noted, $W$ is contained in the Du Val locus. Suppose $D$ has three branches. $\Sigma$ can contain at most one Du Val point (or it will not be contractible) thus $\Sigma$ meets $W$ at two smooth points, and an $A_k$ point. But then

$$1 < 2e + e/(k+1) \leq \Sigma \cdot (cX + dY + fZ) = -\Sigma \cdot K_T \leq 1$$

a contradiction. □

*In view of (11.3.1), we return to the notation of (11.0.1).*

**11.3.2 Lemma.** *If $D$ has two branches at $q$, then one branch has a double point, a simple cusp, and the other is smooth. If $X$ is the branch with the cusp, then $\Sigma$ meets $X$ normally at one smooth point, $\Sigma$ contains two singularities (2) and (3), and $Y$ meets the $-2$-curve, and is disjoint from $\Sigma$ on $\tilde{T}$. In terms of the classification of (11.2), this is configuration III for the genus one cusp $X$.*

*Proof.* Suppose $D$ has two branches. Since $D$ has multiplicity three, one has a double point, say $X$, necessarily a $g^{th}$ order cusp, and the other, $Y$, is smooth. The configuration with respect to the cusp $X$ belongs to the classification at the start of (11.2), by remark (11.2.0).

If $X$ and $Y$ both meet $\Sigma$ at smooth points, then from $e > 1/3$ it follows that each meets $\Sigma$ normally. But then their images at $q$ are analytically isomorphic, a contradiction. So at least one contains an $A_k$ point. If each meets $\Sigma$ on $\tilde{T}$ then

$$(cX + dY) \cdot \Sigma \geq e + e + e/(k+1) > 1$$

a contradiction.

Now consider the configuration for the cusp $X$.

If the configuration is I, $Y$ contains the $A_k$ point and we have

$$1 \geq 2c + d/(k+1) \geq e(2k + 3/k + 1) > 1$$

a contradiction. If the configuration is II, then (since $X$ and $Y$ are disjoint on $T$), $Y$ meets $\Sigma$ normally at a smooth point. We have

$$1 = d + c + c/(k+1) > 1$$

a contradiction.

Since $K_T + X + Y$ is plt, $X + Y$ only contains Du Val points, and $Y$ is smooth at $q$ it follows that the configuration is III, $Y$ meets the $-2$-curve and is disjoint from $\Sigma$ on $\tilde{T}$. Thus $c + \frac{d}{2} + \frac{g}{2g+1} = 1$, and so from the flush condition, $g = 1$. This is the configuration described. □

**11.3.3 Lemma.** *Suppose $Y$ is empty. Then either*

(1) *$\Sigma$ has singularities $(3, 2')$, $(3)$ and meets $X$ normally at a smooth point, or*

(2) *$\Sigma$ has singularities $(3)$, $(2)$ and on the minimal desingularisation $X$ meets $\Sigma$ normally at the intersection of $\Sigma$ and the $-2$-curve.*

*Proof.* Consider the multiplicity, $m$, of the strict transform of $D$ on the first blow up of $q$.

Suppose first $m$ is three. Then by the flush condition, by considering the coefficient of $\Sigma$, $e < 2/5$. Thus, by our simplifying assumptions, $X$ lies in the smooth locus. Also by (10.1) the singularities of $T$ are either Du Val, or type (3). Further, as $e > 1/3$, there can be at most one type (3) singularity, and $\Sigma \cdot X \leq 2$. If $\Sigma$ is tangent to $X$, then $T$ is Du Val and thus by (3.3), $T$ has one Du Val singularity along $\Sigma$. But then $D$ has a double point, as this is configuration $I$ for the cusp, contradiction. So $X$ meets $\Sigma$ normally. $T$ cannot be Du Val or $D$ is smooth. So there is a unique non Du Val singularity along $\Sigma$, a (3) point. By (11.3.4) the singularities along $\Sigma$ are (3) and (2). This is configuration III for a genus one cusp, so $D$ has a double point, a contradiction.

If $m$ is one, then $X$ is triply tangent to $\Sigma^1$ and we must blow up at least twice more along $D$. At this point $T$ has a singularity of type $(\underline{3}')$, and $e = 1/2$, a contradiction. By (11.1), further blow ups will only increase $e$.

Thus $m = 2$. As $D \cdot \Sigma^1 = 3$, $X$ must have an ordinary cusp on $S^1$. The next blow up must be along $X$. This configuration is not allowed, as $c = 2/5$, and $k = 1$. The two subsequent blow ups must be along $X$ and give (2) and (1). On any further blow up there is a singularity of coefficient at least $1/2$, which is not allowed as $e < 1/2$. □

**11.3.4 Lemma.** *If $\Sigma$ contains a unique non Du Val singularity, a (3) point, and $\pi(\Sigma)$ is smooth, then there is exactly one further singularity, an $A_1$ point, along $\Sigma$, that is $T$ is the surface of configuration III for a genus one cusp.*

*Proof.* We consider the sequence of blow ups $h : \tilde{T} \longrightarrow S$. Suppose the first $r \geq 2$ blow ups are over smooth points of $T$. In the resulting configuration the singularity is $(2', \dots, 2) = A_{r-1}$. The next blow up (by the choice of $r$) is at the intersection of $\Sigma$ and the marked $-2$-curve. The resulting configuration has two singularities, $(3', A_{r-2})$ and $(2')$. On any further blow up there is a non Du Val singularity, not of type (3). Thus we have this configuration with $r = 2$. □

**11.4 $K_S + D$ log terminal at $q$.**

Here $Y = \emptyset$, and $D = X$. We make an additional simplifying assumption: *We assume $\tilde{D}^2 > \tilde{W}^2$, or equivalently, that the first blow up in $h : \tilde{T} \longrightarrow \tilde{S}$ is at the point of $\tilde{D} \subset \tilde{S}$ lying over $q$.* Note this holds for $W = E_i$ in a step of the hunt, since (in the hunt notation) $\tilde{D}_i$ is $K$-negative, while $\tilde{E}_i$ is $K$ non-negative.

There are infinitely many possible configurations, we now indicate how they can be generated. *We will not assume in the analysis below that $K_T + cX$ is either flush or $\pi$-trivial. We will only use our assumption on $\tilde{D}^2$, and that $K_T + W$ and $K_S + D$ are log terminal*

Consider the possibilities for $h : \tilde{T} \longrightarrow \tilde{S}$. By assumption, the first blow up of $h : \tilde{T} \longrightarrow \tilde{S}$ is along $X = D$. Subsequent blow ups are of three sorts. Either we blow up at the point of $\Sigma \subset \tilde{T}$ lying over $\Sigma \cap X$, which we call the **nearest** point, or at a point, lying over a singular point of $T$, different from the near point, which we call the **farthest** point (there will be at most one such point), or at a point on $\Sigma$, not along $X$, lying over a smooth point of $T$. We call this last case an **interior blow up**. We indicate a series of non-interior blow ups by a sequence, $(n; f; n^2)$ for example, where $n$ indicates a blow up at the nearest point to $X$, and $f$ indicates a blow up at the farthest point. We count the first blow up of $h$ as a blow up at the nearest point. We will indicate an interior blow up by $i$. After an interior blow up, there is only one singular point of $T$ on $\Sigma$. A further interior blow up is impossible, for then $K_T + W$ is not lt, and so the possible sequence of blow ups is unique. We write $n^k$ if there are $k$ further blow ups. Note $K_T + X + \Sigma$ is lc iff there is no interior blow up, and that $X \cap \Sigma$ is a smooth point of $T$ iff all the blow ups are at the near point, that is iff the configuration is $(n^r)$ for some $r \geq 0$.

For example if the singularity at $q$ is $(\underline{2}, 3)$, and we make the blow up sequence $(n^2; f; i; n)$, the singularities on $T$ along $\Sigma$ are $(\underline{2}, 3', 3, 3, 3)$ and $(2')$.

Here is a picture:

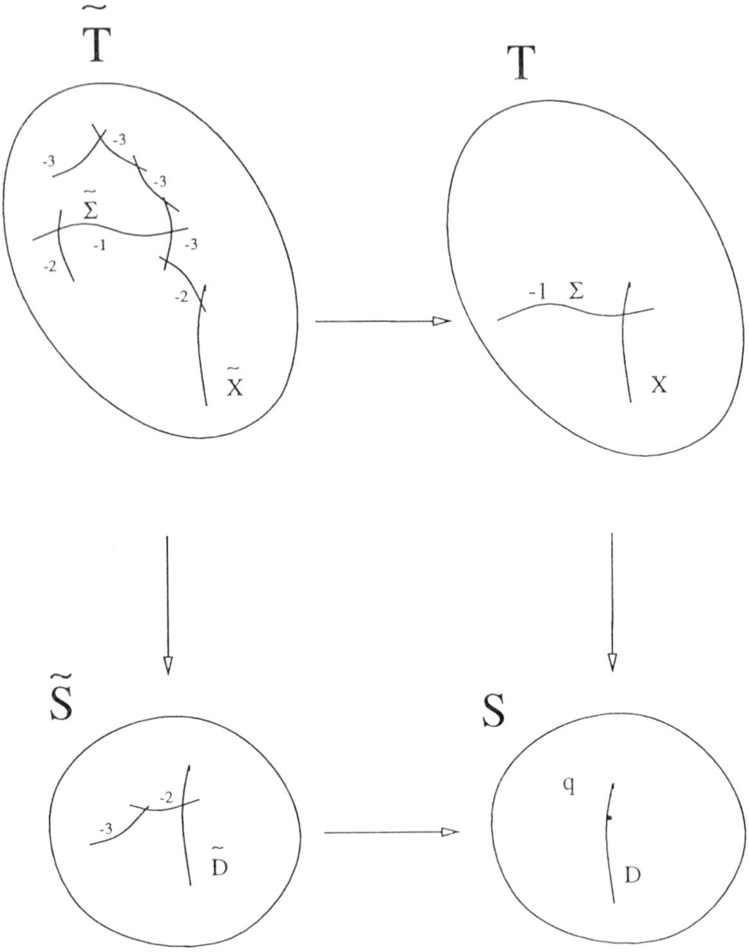


Figure 7

**11.5 $\mathbb{P}^1$-fibrations.**

*11.5.1 Notation: Let $(p, C)$ be a smooth curve germ, and let $\pi : T \longrightarrow C$ be a (proper) $\mathbb{P}^1$-fibration of relative Picard number one, with $T$ a normal surface. Let $\tilde{\pi}$ be the composition*

$$\tilde{\pi} : \tilde{T} \to T \xrightarrow{\pi} C$$

*(where, as throughout the paper, $\tilde{T} \longrightarrow T$ is the minimal desingularisation). Let $\mathcal{F} \subset \tilde{T}$, $\mathcal{G} \subset T$ be the scheme-theoretic fibres (over $p$), with reductions $F, G$.*

**11.5.2 Lemma.** *The following are equivalent:*

(1) $\pi$ *is smooth.*

(2) $T$ *is smooth.*

(3) $\mathcal{G}$ *is generically reduced.*

*Proof.* The equivalence of (1-2) is well known, and obviously (1) implies (3). Assume (3). $\mathcal{G} \subset T$ is Cartier, and hence Cohen-Macaulay. Thus it has no embedded points, and so is integral. It follows that

$$h^1(\mathcal{O}_G) = 1 - \chi(\mathcal{O}_G) = 1 - \chi(\mathcal{O}_{\mathbb{P}^1}) = 0$$

thus $G = \mathbb{P}^1$. Since $\mathcal{G}$ is Cartier, (2) follows. □

*From now on we assume $T$ is singular.* In view of (11.5.2) we will sometimes abuse notation and call $G$ a *multiple* or *singular* fibre of $T$ (even though $G$ is by assumption reduced, and is isomorphic to $\mathbb{P}^1$ ).

**11.5.3 Lemma.** $\tilde{G} \subset F$ *is the unique* $-1$*-curve in the fibre.*

*Proof.* This holds since the exceptional locus of $\tilde{T} \longrightarrow T$ is $K_{\tilde{T}}$ nef. □

*11.5.4 Further Notation: Let*

$$\begin{array}{ccc} \tilde{T} & \xrightarrow{h} & W \\ {\scriptstyle \tilde{p}}\downarrow & & \downarrow{\scriptstyle f} \\ C & = & C \end{array}$$

*be a relative minimal model over $p$, so $f$ is a smooth $\mathbb{P}^1$-bundle, and $h$ is a composition of smooth blow ups. Let $H$ be the fibre of $f$ over $p$. In the case where the dual graph of the inverse image of $F$ is a chain, we will use the following notation to describe the scheme-theoretic fibre $\mathcal{F}$: The sequence $k(-a) + l(-b) + \cdots + m(-c)$, is taken to mean that the first rational curve in the chain has self-intersection $-a$ and has multiplicity $k$ inside the fibre, the second has self-intersection $-b$ and multiplicity $l$ and so on.*

Consider the possibilities for $h$. The first blow up is at some point along the fibre, after which the fibre is $(-1) + (-1)$. One of the components is (the strict transform of) $H$, the other, $V$, is $h$-exceptional.

There must be a further blow up. Necessarily, by (11.5.3), at the intersection point of the two $-1$-curves, after which the fibre is $(-2) + 2(-1) + (-2)$. All subsequent blow ups are at some point along the unique $-1$-curve, by (11.5.3). At each stage one either blows up at a point where the $-1$-curve meets another fibral component, or at a point of the $-1$-curve which is not on any other component. We call the latter an **interior blow up**. Note in particular that whatever the sequence of blow ups there are exactly two reduced components in the fibre, the strict transforms of $V$ and $H$. We collect these observations as:

**11.5.5 Lemma.** *Notation as in (11.5.1), (11.5.4). $\mathcal{F}$ has exactly two reduced components. The collection of multiple components is contractible. Let $h' : T \longrightarrow W'$ contract these components. $W'$ is smooth, and $f' : W' \longrightarrow C$ has fibre $(-1)+(-1)$. $h' : T \longrightarrow W'$ blows down (at each stage) the unique $-1$-curve. $f'$ has two relative minimal models $g_1$, $g_2$ given by contracting one, or the other $-1$-curve. $\tilde{\pi}$ has exactly two relative minimal models $g_1 \circ h'$ and $g_2 \circ h'$, and the relative minimal model is determined by choosing which reduced component of $F$ not to contract. This component is the strict transform of $H$, the other component is the strict transform of $V$. $T$ has rational singularities, and at most two singular points along $G$.*

*Proof.* The last two statements are the only new observations, and follow easily from the description of the morphism $h'$. □

Possible $\pi$ are classified by the choices of the blow ups in $h$. Using the classification of lt singularities, one can further classify possibilities when $T$ is log terminal. One easily checks in particular the following:

**11.5.6 Lemma.** *Notation as in (11.5.1). $K_T + G$ is log terminal iff there are no interior blow ups.*

Note after the first interior blow up, $F$ consists of the $-1$-curve, and a chain of $K_T$-positive curves. We call this chain (and its strict transform after further blow ups), the **principal** chain of the fibre. Its two ends are the two reduced components of the fibre.

**11.5.7 Fibrations with a section.**
*Additional Notation: Let $E \subset T$ be a reduced curve germ, finite over $C$, such that $K_T + E$ is lt, and $\tilde{E} \cdot \mathcal{F} = 1$.*

Necessarily $E$ is a section of $\pi$, $\tilde{E}$ meets a unique components of $\mathcal{F}$, a reduced component, and $\tilde{E}$ has normal contact. By (11.5.5) there is a unique choice of $h$ which is an isomorphism in a neighbourhood of $\tilde{E}$. There can be at most one interior blow up (otherwise $E$ contains a non-chain singularity). Let $x = E \cap G \in T$. If $K_T + G$ is lt then $F$ (and hence $h$) is uniquely determined by the marked singularity $(x, E)$, and otherwise it is determined by the marked singularity $(x, E + G)$. We indicate the singularities as in (11.0.1), with an underline (resp. prime) indicating normal contact with $\tilde{E}$ (resp. with $\tilde{G}$).

*Remark.* Of course one can classify global $\mathbb{P}^1$-fibrations $\pi : T \longrightarrow C$, for a smooth proper curve $C$, by using the above local classification, so long as one has a classification of smooth $\mathbb{P}^1$-bundles $W \longrightarrow C$. For example one can carry this out for $C = \mathbb{P}^1$.

*11.5.8 Example.* Suppose $C = \mathbb{P}^1$, $E \subset T$ is a $-k$-curve, and a section, with $K_T + E$ log terminal, and $E$ contains a single singularity $x = (\underline{2}, 2, 2)$. Let $G$ be the (reduction of the) fibre through $x$.

Then the above analysis shows: $\pi$ is smooth away from $G$. There is a unique relative minimal model $h : \tilde{T} \longrightarrow W$ which is an isomorphism in a neighbourhood of $\tilde{E} \subset \tilde{T}$. $W = \mathbb{F}_k$ and (the image of) $E \subset W$ is the negative section. If $K_T + G$ is log terminal, then $F$ is $(-\underline{2}) + 2(-2) + 3(-2) + 4(-1) + (-4)$. If $K_T + G$ is not lt, then $x$ is the principal chain, the only singularity along $G$, and the marked fibre is $(-\underline{2}) + 2(-2') + (-2)$.

Here is a picture:

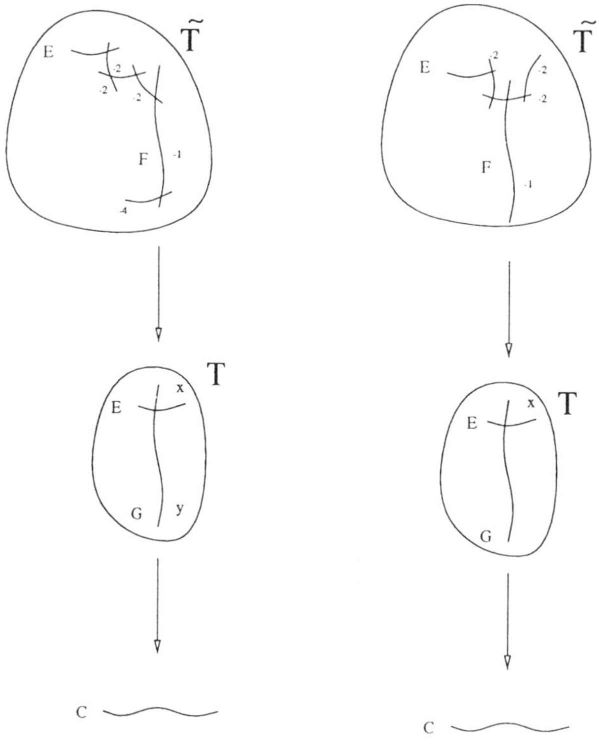


Figure 8

**11.5.9 Lemma.** *Notation as in (11.5.1). Assume $T$ is log terminal, $G$ is a multiple fibre of $\pi$, of multiplicity $m$, and $G$ contains a cyclic singularity, either Du Val or almost Du Val. If $e(T) < 2/3$, then $G$ is one of the following:*

*If $K_T + G$ is not lt at any singular point:*

(1) $(2, 2', 2)$, $m = 2$ *or*

(2) $(3,2',2,2)$, $m=3$.

*If $K_T+G$ is log terminal at one singular point, but not lt:*

(3) $(2';z)$, $m=4$. *$z$ is a non-chain singularity, with centre $-2$ and branches $(2)$, $(2)$, and $(2,\dots,3')$ (or $(3')$).*

(4) $(2,3',2;2')$ $m=4$.

*If $K_T+G$ is log terminal:-*

(5) $(A_k;(k+1)')$, $k\le 4$, $m=k+1$. *The fibre is*

$$-(k+1)+[k+1](-1)+k(-2)+[k-1](-2)+\cdots+(-2).$$

(6) $(2,3';2',3)$, $m=5$. *The fibre is* $(-2)+2(-3)+5(-1)+3(-2)+(-3)$.

(7) $(3,2,2';4',2)$ $m=7$. *The fibre is* $(-3)+3(-2)+5(-2)+7(-1)+2(-4)+(-2)$.

(8) $(4,2';3',2,2)$, $m=7$. *The fibre is* $(-4)+4(-2)+3(-1)+7(-3)+2(-2)+(-2)$.

*Proof.* One checks that each of (1-8) is an allowed configuration. Suppose $K_T+G$ is lt. One checks that for each configuration (5-8), and either of the two possible non-interior blow ups, the resulting configuration is either one of (5-8), or disallowed by the hypotheses. Since the fibre $(-2)+2(-1)+(-2)$ is (5), $k=1$, we have considered all possibilities.

Now suppose $K+G$ is not lt. Thus there is some interior blow up. Consider the sequence of blow ups. We begin with a sequence of non-interior blow ups, giving one of the fibres (5-8). Call this configuration $J$. Then we make a string $i^r$ of $r\ge 1$ interior blow ups.

Suppose there are two singular points along $G$. Then we make in addition at least one non-interior blow up. $e(T)\ge 2/3$ unless $J$ is given by (5) with $k=1$. $r\ge 2$ gives (3), and $r=1$ gives (4). Any further blow up has $e(T)\ge 2/3$.

Suppose there is only one singular point along $G$, cyclic and Du Val or almost Du Val by assumption. Then $r=1$ (otherwise the singularity cannot be cyclic). $J$ given by (5) with $k=1,2$ are the only possibilities in which the singularity is Du Val or almost Du Val. This gives (1) and (2). No further blow up is possible (the next blow up has to be non-interior), as the singularity will not be Du Val or almost Du Val. □

**11.5.10 Lemma.** *Suppose $D\subset\mathbb{F}_n$ is an irreducible curve of degree $d\ge 2$, arithmetic genus $g$. If $g\le 1$, then either*

(1) $g=0$, $n=0$ *and* $D^2=2d$, *or*

(2) $g=0$, $d=2$, $n=1$ *and* $D^2=4$, *or*

(3) $g=1$, $d=3$ *and* $n=1$, *or*

(4) $g=1$, $d=2$ *and* $n\le 2$.

*Proof.* $(K_S+D)\cdot D \leq 0$. But $(K_S+D)\cdot F \geq 0$, where $F$ is a smooth fibre, and $(K_S+D)\cdot E_\infty > 0$, unless $n \leq 2$, where $E_\infty$ generates the other extremal ray. Write $D$ numerically as $dE + rF$. One computes $r$ using $(K + D) \cdot D = 2g - 2$, and then checks the result. □

**11.5.11 Lemma.** *Suppose $\pi : T \longrightarrow \mathbb{P}^1$ is a log Fano fibration of relative Picard number one, general fibre $\mathbb{P}^1$, $C$ is a section, $D$ is rational, an irreducible double section, and $C$ and $D$ meet at only smooth points of $T$. Let $F_1$ and $F_2$ be the fibres, where $\pi|_D$ has a ramification point. Let $W$ be the unique minimal model $h : \tilde{T} \longrightarrow W = \mathbb{F}_n$, of (11.5.5), (11.5.7), such that $h$ is an isomorphism in a neighbourhood of $\tilde{C}$. If $K_T + D$ is lt, then*

(1) *Any multiple fibre is either $F_1$ or $F_2$ (though either $F_i$ might be smooth).*
(2) *Neither $C$ nor $D$ contains more than two singularities of $T$.*
(3) *$h|_D$ is an isomorphism.*

*Proof.* Let $G$ be a multiple fibre. Suppose $G$ is neither $F_1$ nor $F_2$. Then each point of $G \cap C$ and $G \cap D$ is a singular point, contradicting (11.5.5). Hence (1).

(2) follows easily, using (11.5.5).

Suppose (3) does not hold. Then one of the ramification points of $D$ is smooth on $T$, and has a cusp on $W$. Considering the first two blow ups of $h$, we see that $D$ does not meet the exceptional locus of $\tilde{T} \longrightarrow T$ transversally, a contradiction (since we assume $K_T + D$ is lt). □

**11.5.12 Lemma.** *Notation as in (11.5.1), (11.5.4). Suppose $\pi : T \longrightarrow C$ is a $\mathbb{P}^1$-fibration, and $G \subset T$ is an irreducible fibre contained in the Du Val locus of $T$. One of the following holds:*

(1) *$T$ is smooth along $G$.*
(2) *There are exactly two singularities, $A_1$ points, along $G$. $K_T + G$ is lt. The fibre is $(-2) + 2(-1) + (-2)$ (notation as in (11.5.4)).*
(3) *There is a unique singularity, an $A_3$ point along $G$, $(2, 2', 2)$.*
(4) *There is a unique singularity, a $D_n$ point, along $F$, its branches are $(2)$, $(2)$ and $(\underline{2}, \dots, 2')$, where the central curve meets the underlined curve, and $G$ meets the primed curve.*

*Proof.* The only possibilities for $h$ which yield Du Val singularities are $(-2) + 2(-1) + (-2)$, or this configuration followed by a sequence of interior blow ups. □

We will use the following ad-hoc result in §14.

**11.5.13 Lemma.** *Notation as in (11.5.1). If $G$ is a multiple fibre of multiplicity $m = 3$, and the coefficient $e(T) < 2/3$ then $G$ is one of the fibres in the classification (11.5.9).*

*Proof.* Consider the possibilities for $h$ as below (11.5.4). It is easy to see that the only way to get a fibre of multiplicity three is to begin with $(-3) + 3(-1) + 2(-2) + (-2)$ and make a series of $r \geq 0$ interior blow ups. After two interior blow ups there is a non-chain point of coefficient $2/3$. For $r \leq 1$, the fibres are in (11.5.9). □

## §12 THE LINEAR SYSTEM $|K_S + A|$

**The following notation is fixed throughout the section.** *Let $S$ be a rank one log del Pezzo surface. Let $A \subset S$ a reduced and irreducible divisor, such that $-(K_S + (1/2)A)$ is ample and $K_S + A$ is lt at singular points of $S$. Note in particular $A$ has planar singularities, and is thus Gorenstein. Assume the arithmetic genus $g$ of $A$ is at least one. $r : \tilde{S} \longrightarrow S$ denotes the minimal desingularisation of $S$, and tildes denote strict transforms. Let $D = F + M$ of (12.0) below. Let $\overline{D}$ be the reduction of $D$. Let $Y \longrightarrow S$ extract exactly the exceptional divisors (of the minimal desingularisation) adjacent to $A$.*

Note the assumptions imply $r|_{\tilde{A}}$ is an isomorphism.

**12.0 Definition-Lemma.** *$h^1(\omega_{\tilde{S}}) = h^0(\omega_{\tilde{S}}) = 0$ and $H^0(\omega_{\tilde{S}}(\tilde{A})) \simeq H^0(\omega_A)$. There are integral, effective, Weil divisors $F, M \subset S$, $E \subset \tilde{S}$ with the following properties:*

(1) $\tilde{F} + \tilde{M} + E \in |K_{\tilde{S}} + \tilde{A}|$, $F + M \in |K_S + A|$.
(2) *The support of $E$ is $\tilde{S} \longrightarrow S$ exceptional.*
(3) $|\tilde{F}|$ *has no fixed components. No component of $\tilde{M}$ moves.*

*Proof.* The first equalities hold since $\tilde{S}$ is rational. The second follow from the first and the exact sequence

$$0 \longrightarrow \omega_{\tilde{S}} \longrightarrow \omega_{\tilde{S}}(\tilde{A}) \longrightarrow \omega_A \longrightarrow 0.$$

By assumption, $|\omega_A|$ is non-empty. Thus $|K_{\tilde{S}} + \tilde{A}|$ is non-empty. We can write a general member as $\tilde{F} + \tilde{M} + E$, where $\tilde{F}$ is the moving part, $\tilde{M} + E$ is the fixed part, and $E$ is the $\tilde{S} \longrightarrow S$ exceptional part. Now let $F, M$ be the images of $\tilde{F}, \tilde{M}$. Note $\tilde{F}$ and $\tilde{M}$ are indeed the strict transforms of $F$ and $M$, by construction. □

**12.1 Lemma.** *Notation as in (12.0).*

(1) *$K_S + D$ is negative (that is anti-ample) and $\overline{D}$ has arithmetic genus zero.*
(2) *If $D$ is reducible, then the union of the pairwise intersection of components of $D$ consists of a single point $b$.*
(3) *$F \neq \emptyset$ iff $g \geq 2$. $|\tilde{F}|$ is basepoint free.*
(4) *$\overline{D} \cap \operatorname{Sing}(S)$ contains any non Du Val singularity of $S$, and any singularity of $S$ which lies on $A$.*
(5) *If $\overline{D}$ is reducible, then $K_S + \overline{D}$ is plt away from $b$.*
(6) *$\tilde{A}$ is disjoint from $\tilde{M}$ and $E$.*

(7) *The support of* $\tilde{M}$ *is a disjoint union of* $-1$*-curves.* $\tilde{M} \subset Y$ *is contractible by a series of* $K$*-negative contractions, and the number of irreducible components of* $M$ *is at most the number of singularities along* $A$.

*Proof.* $K_S + D = 2(K_S + 1/2A)$ is negative. (1), (2) and (5) follow by adjunction.

$h^0(\omega_A) \geq 1$ iff $g(A) \geq 2$. (3) follows, since a moving smooth rational curve on a surface with $H^1(\mathcal{O}_S) = 0$ is basepoint free.

(6) follows from (12.7).

From $K_{\tilde{S}} + \tilde{A} - \tilde{M} - \tilde{F} = E$, we see that $E$ is nef, and hence by (2.19) of [27], empty, over any point not contained in $M + F$. (4) follows.

For (7) we use the following:

**Claim.** *Let* $\gamma : Y \longrightarrow Z$ *be proper birational map, which is an isomorphism in a neighbourhood of* $\tilde{A} \subset Y$. *Let* $N$ *be an effective divisor on* $Y$, *disjoint from* $\tilde{A}$, *such that* $\gamma$ *and* $Y \longrightarrow S$ *are finite on* $N$. *Then*

$$0 > \gamma(N) \cdot (K_Z + \gamma(\tilde{A})) = K_Z \cdot \gamma(N).$$

*Proof.* Since $-K_S$ is ample, there is an effective $\mathbb{Q}$-divisor $J$ supported on the exceptional locus of $Y \longrightarrow S$ such that $-(K_Y + J)$ is ample. We have

$$0 > (K_Z + \gamma(J)) \cdot \gamma(N) \geq K_Z \cdot \gamma(N) = (K_Z + \gamma(\tilde{A})) \cdot \gamma(N). \quad \square$$

Now let $\gamma : Y \longrightarrow Z$ contract some components of $\tilde{M}$ (we allow $\gamma$ to be the identity). Let $N$ be the reduction of some component of $M$ which is not contracted. By (6) and the claim

$$0 > \gamma(N) \cdot (K_Z + \gamma(\tilde{A})) = \gamma(N) \cdot K_Z = \gamma(N) \cdot \gamma(\tilde{M} + \tilde{F} + E) \geq \gamma(N)^2$$

Thus $\gamma(N)$ can be contracted, in fact by a $K_Z$-negative contraction. Thus, $M$ can be contracted by a series of $K$-negative contractions. (7) follows. $\square$

**12.2 Lemma.** *Suppose* $S = S_1$ *of the hunt. If* $S_0$ *does not have a tiger, then every component of* $M$ *contains at least two singularities of* $S$.

*Proof.* Suppose $N$ is a component of $M$ passing through a unique singular point $z$ (it must contain a singular point, or it would be contractible on $S$). As $\tilde{M}$ is disjoint from $\tilde{A}$, by (12.1.6), $z$ is a point of $A$. $M$ cannot meet $\Sigma_1$ on $T_1$, for otherwise $z = \pi(\Sigma_1)$, a singular point of $A$. Thus $M$ meets $E_1$ at a singular point of $T_1$ and $x_0$ is the only singular point of $S_0$ along $M$.

Let $D$ be the unique exceptional curve of $\tilde{S}_0$ which meets $M$ (there is a unique $D$ by (4.12.2) and (12.1.1)), and let $g : T' \longrightarrow S_0$ be the extraction of $D$. Then $M$ is a $-1$-curve in the smooth locus of $T'$, and we may contract $M$ on $T'$ to obtain $S'$. Let $D'$ be the image of $D$ in $S'$.

By the classification of lt points, if $K_{S'} + D'$ is lc then it is anti-nef and we have a tiger. So we can assume that $x_0$ is a non-chain singularity and one of the singular points along $E_1$ other than $z$ has index at least three. Thus $z$ has index at most 5. As $M^2 > 0$, it follows that $z = A_4$ (just consider the possibilities, see for example the proof of (10.6)), and $D$ is the primed curve $(2, 2', 2, 2)$. But then $(K_S + A) \cdot M \leq -1 + 3/5 < 0$, and $-(K_S + A)$ is ample, a contradiction. □

**12.3.** *Notation as in (12.0). Suppose $g = 2$. Then*

(1) *$\tilde{F}$ is the fibre of a ruling on $\tilde{S}$, $\tilde{A}$ is a double section.*
(2) *$\tilde{F}$ is disjoint from $\tilde{M}$ and $E$ on $\tilde{S}$.*
(3) *If $p \in F$ is a singular point, then either $F$ meets the same curve over $p$ as $A$, or $F$ meets a curve of self-intersection at most $-3$.*

*If $p \in F \setminus F \cap M$ is a singular point, then $E$ is empty over $p$ and*

(4) *If $p \in A$ then $p$ is Du Val and $F$ meets the same exceptional curve at $p$ as $A$.*
(5) *If $p \notin A$ then $p$ has weight one, and $F$ meets the unique $-3$-curve.*

*Proof.* By (12.1.3) and (12.1.6) we have

$$\begin{aligned} 0 \leq \tilde{F}^2 \leq (\tilde{F} + \tilde{M} + E) \cdot \tilde{F} = (K_{\tilde{S}} + \tilde{A}) \cdot \tilde{F} &\leq -2 + \tilde{A} \cdot \tilde{F} \\ &= -2 + \tilde{A} \cdot (K_{\tilde{S}} + \tilde{A}) = 2(g-2). \end{aligned}$$

(1-2) follow.

For (4) and (5), (2) and the equality $K_{\tilde{S}} + \tilde{A} - \tilde{F} = E$ over $p$ imply $E$ is nef over $p$, thus empty, by (2.19) of [27]. (4) and (5) follow. (3) is similar. □

**12.4 Lemma-Definition.** *Notation as in (12.0). Suppose $g = 1$. By (12.1.7) $\tilde{M} \subset Y$ is contractible, let $\pi : Y \longrightarrow W$ be the contraction. Let $G = \sum G_i$ be the exceptional locus of $f : Y \longrightarrow S$. By the definition of $Y$, $G$ is the union of the exceptional divisors (of $\tilde{S}$) adjacent to $A$. Let $D_i = \pi(G_i)$. Let $\tilde{A} \subset Y$ be the strict transform of $A$. We have*

(1) *The rank of $W$ is one more than the difference between the number of singular points on $A$ and the number of irreducible components of $M$.*
(2) *$\pi$ is an isomorphism in a neighbourhood of $\tilde{A}$.*
(3) *$A \in |-K_W|$, $A \subset W^0$, and $W$ is Gorenstein.*
(4) *$D_i \cdot A = 1$.*
(5) *If $A$ contains only one singular point, then $W$ has rank one and $\tilde{A}^2 \geq 1$.*

*Proof.* The relative Picard number of $f : Y \longrightarrow S$ is the number of singularities along $A$, and to get from $Y$ to $W$, we contract $M$. Hence (1).

(2) is immediate from (12.1.6).

By (12.1.3) and (12.0.1), $K_Y + \tilde{A} = \tilde{M}$. (3) follows from this, and (2).

(4) follows from (3) since $G_i \cdot \tilde{A} = 1$, as $K_S + A$ is lt at singular points

$\tilde{A}^2 = K_W^2$, by (2) and (3). If $W$ has rank one, $K_W^2 > 0$. Hence (5). □

*Remark 12.4.6.* Suppose $S = S_1$, $A = A_1$ of the hunt, and $g(A) = 1$. The transformation $S_1 \dashrightarrow W$ of (12.4) is not in general the next hunt step. However, since $W$ is Gorenstein, it is often a useful alternative, especially if one can argue that $W$ has rank one.

**12.5 Lemma.** *Notation as in (12.0). If $A$ has genus one and $K_S + 2/3A$ is negative then no irreducible component of $M$ passes through more than two singular points of $S$.*

*Proof.* Suppose not. Let $N$ be the reduction of an irreducible component of $M$ containing three singular points, say of index $p$, $q$ and $r$. As $K_S + (2/3)A$ is negative, $K_S + 2N$ is negative. But by adjunction

$$\begin{aligned}(K_S + 2N) \cdot N &\geq (p-1)/p + (q-1)/q + (r-1)/r - 2 + N^2 \\ &\geq (p-1)/p + (q-1)/q + (r-1)/r - 2 + (-1 + 1/p + 1/q + 1/r) = 0,\end{aligned}$$

a contradiction. □

**12.6 Canonical map of an integral Gorenstein curve.**

Here we prove the following, as we were unable to find it in the literature:

**12.7 Lemma.** *Let $A$ be a proper, integral, Gorenstein curve. If $A \neq \mathbb{P}^1$ then $|\omega_A|$ is basepoint free.*

**12.8 Lemma.** *Let $A$ be a proper, integral, Gorenstein curve. There is a differential form $F \in H^0(\omega_A)$ which does not vanish at any singular point of $A$.*

*Proof.* Let $\pi : \tilde{A} \longrightarrow A$ be its normalisation. Define the Cartier divisor $D$ by $\omega_{\tilde{A}}(D) = \pi^*(\omega_A)$ (thus $D$ is cut out by the conductor ideal on $\tilde{A}$). We have a commutative diagram with exact rows

$$\begin{array}{ccccccccc} 0 & \longrightarrow & \pi_*(\omega_{\tilde{A}}) & \longrightarrow & \pi_*(\omega_{\tilde{A}}(D)) & \longrightarrow & \pi_*(\omega_D) & \longrightarrow & 0 \\ & & \downarrow & & \| & & \downarrow{\scriptstyle g} & & \\ 0 & \longrightarrow & \omega_A & \longrightarrow & \omega_A \otimes \mathcal{O}_{\tilde{A}} & \longrightarrow & \omega_A \otimes (\mathcal{O}_{\tilde{A}})/(\mathcal{O}_A) & \longrightarrow & 0. \end{array}$$

This induces a short exact sequence

$$0 \to H^0(\omega_{\tilde{A}}) \to H^0(\omega_A) \to \ker(g) \to 0.$$

Fix a singular point $q \in A$ and a local generator $F_q \in \omega_{A,q}$. Fix a neighbourhood $q \in U$ missing the other singularities, and let $V = \pi^{-1}(U)$. The polar divisor of $F_q$, thought of as a rational form on $V$, is exactly $D|V$, and the image of $F_q$ in $\omega_D|V$ is an element $G_q \in \ker(g)$. Any lifting of $G_q$ to $H^0(\omega_A)$ has polar divisor $D|V$ on $V$, and so is a generator for $\omega_{A,q}$. □

*Proof of (12.7).* By (12.8) we need only show there are no smooth basepoints. By Duality $p$ is not a basepoint for $|\omega_A|$ iff $\mathrm{Hom}(m_p, \mathcal{O}_A)$ is one dimensional. If $p$ is a smooth point $\mathrm{Hom}(m_p, \mathcal{O}_A) = H^0(\mathcal{O}(p))$, so if the dimension is at least two, then $|p|$ is basepoint free, and defines a birational map to $\mathbb{P}^1$, and so an isomorphism. □

## §13 Classification of Bananas and Fences

**13.0 Notation, fixed throughout §13:** *$S$ is a rank one log del Pezzo surface. $X, Y \subset S$ are irreducible rational curves such that $K_S + X + Y$ is log terminal at singular points of $S$.*

For the definitions of Fence and Banana see (8.0.11-12).

**13.1 Lemma.** *Notation as in (13.0). If $X$ and $Y$ are smooth then each of $X$, $Y$ contains at most one singular point of $S$. If furthermore there is a singularity off $X \cup Y$, then $S = \overline{\mathbb{F}}_n$, and $X, Y \in |\sigma_n|$ (so in particular, $X + Y \subset S^0$).*

*Proof.* We argue first that each of $X, Y$ contains at most one singular point.

Let $B$ be either $X$ or $Y$. By (9.2), $S \setminus B$ contains either a unique singularity, or two $A_1$ points. Since there is no $S(4A_1)$ in the list (3.1), there is at least one component which contains a unique singular point.

Thus we can (switching if necessary) assume $X$ contains a unique singular point, and (arguing by contradiction), $Y$ contains exactly two singular points, each $A_1$ points. Note $K_S + X$ and $K_S + Y$ are both anti-ample, by adjunction.

Consider the hunt with respect to $K_S + X$, which is clearly flush. Obviously $x$, for the first hunt step, is the unique singularity along $X$. Note that after this we will not use flushness.

Note $X \subset T^0$.

Suppose $T$ is a net. If $X$ is fibral, then it is a smooth fibre and by negativity $Y$ is a section, which meets $X$ once. But then $T$ has two fibres which both contain one $A_1$ singularity, impossible. Thus, since $K_S + X$ is negative, $X$ is a section, in the smooth locus of $T$. Thus $T$ is smooth, a contradiction (there are two singularities along $Y$).

Hence we have a contraction $\pi : T \longrightarrow S_1$. Since $X \subset T^0$, and $K_S + X$ is negative, $\Sigma \cap X = \emptyset$, and $X \subset S_1^0$. Note $K_{S_1} + Y$ is negative, so $\Sigma \cap Y$ cannot contain both of the $A_1$ points (otherwise $Y \subset S_1$ has a singularity at $q$). Thus by (13.7), $S_1 = \overline{\mathbb{F}}_2$, $C + X \subset S_1^0$, $X$ and $C$ are linearly equivalent and $X^2 = C^2 = 2$, where $C = \pi(E)$. A contradiction, since $X \cdot C = 1$.

So each of $X$, $Y$ contains at most one singular point of $S$. Now suppose there is a singularity $z$ off $X + Y$, and a singularity $w$ on $X + Y$. Then by (9.2), both of $z$ and $w$ are $A_1$ points. Thus

$S$ has at least two singular points, and all of its singularities are $A_1$ points. This contradicts the list (3.1). Now apply (13.7). □

**13.2 Proposition (The Smooth Banana).** *If $K_S + X + Y$ is a banana, with $X$ and $Y$ smooth, and $S$ is singular at some point of $X \cup Y$, then each of $X$ and $Y$ contains a unique singular point. Moreover one component is a 2-curve and contains an $(r+1, 2)$ point, the other is a 1-curve and contains an $A_r$ singularity. After extracting the $-(r+1)$-curve, $E$, there is a $K$-negative contraction down to $\overline{\mathbb{F}}_2$, and the image of $X + Y + E$ is two sections (in $|\sigma_2|$) and a fibre. After extracting the adjacent $-2$-curve, $E'$, of the $A_r$ point, there is a $K$-negative contraction down to $\mathbb{P}^2$, and the image of the (component of $X + Y$ which is a) 2-curve is a conic, the image of the 1-curve is a secant line to the conic, and the image of $E'$ is a tangent line to the conic.*

*Moreover if $K_S + aX + bY$ is flush and anti-ample and does not have a tiger, then the next hunt step for $K_S + aX + bY$ extracts $E$, and $S_1 = \overline{\mathbb{F}}_2$.*

*Proof.* Suppose $Y$ is contained in the smooth locus. Then $(K_S+X+Y)\cdot Y = 0$, and $K_S+X+Y$ is numerically trivial. But then $(K_S + X + Y) \cdot X = 0$ which implies $X \subset S^0$, a contradiction. Of course the same applies to $X$, thus either component contains a singular point. By (13.1) each contains a unique singular point, and $S$ is smooth off $X + Y$.

Switching components if necessary, we can assume $X$ contains the singular point of highest index. Note $-(K_S + X)$ and $-(K_S + Y)$ are both ample, by adjunction.

Consider the hunt with respect to $K_S + X$. Obviously $x$, for the first hunt step, is the unique singularity along $X$. We note that after scaling some coefficient could be greater than one (since we are not assuming there is no tiger). This will not effect our argument, which will not use flushness at all.

$T$ cannot be a net, as in the proof of (13.1).

Hence we have a contraction $\pi : T \longrightarrow S_1$. Since $X \subset T^0$, and $K_S+X$ is negative, $\Sigma \cap X = \emptyset$, and $X \subset S_1^0$. Then by adjunction $K_{S_1} + X + Y$ is numerically trivial. Thus $X + Y \subset S_1^0$, and $S_1$ is Du Val. By (13.7) $S_1$ is either $\mathbb{P}^2$ or $\overline{\mathbb{F}}_2$.

Let $C \subset S_1$ be the image of $E$.

If $S_1 = \overline{\mathbb{F}}_2$, $C$ is a fibre, thus $C$ contains a singular point, and so (since $E$ contains at most one singular point), $\Sigma$ meets $E$ at a smooth point, and $Y$ at a singular point. $Y + C$ has an ordinary node at $q$. The given description now follows from (11.1.0).

If $S_1 = \mathbb{P}^2$, then $X$ and $C$ are lines and $Y$ a conic tangent to $C$ at $q$. Now consider the composition of blow ups $h : \tilde{T} \longrightarrow S_1$, as in §11. The first two blow ups must be at $Y$ (to separate $Y$ and $C$). In this configuration, $C$ is a $-1$-curve. Since $\tilde{C}^2 \leq -2$, the next blow up is also at $C$. Now there is only one possible choice for any further blow up such that $K_T + Y$ is plt. This yields the given description.

Now suppose $K_S + aX + bY$ is flush and anti-ample, we do not have a tiger, and the next hunt step yields $\mathbb{P}^2$. We may assume (by switching components if necessary) that $x \in X$. As we get a tacnode, $e < 2/3$ by (11.1.1). Then by (8.0.7.2), the spectral value of $(r+1, 2)$ is at most one, thus $r = 1$, $x$ is an $A_1$ point and the singularity on $Y$ is an $A_2$ point. By adjunction $b > 3/4$ (or $K_S + X + bY$ is anti-nef) and so, by flushness, $a/2 = e > (3/4)(2/3) = 1/2$. Thus $a > 1$, a contradiction. □

**13.3 Proposition (Fence with a smooth side).** *Let $S$ be a log del Pezzo surface of rank one and suppose $X + Y$ is a fence, with $Y$ smooth.*

*If $X$ is smooth, then $-(K_S + X + Y)$ is ample, and $S$ is smooth away from $X + Y$.*

*If $X$ is singular, then $Y$ is a $-1$-curve and contains at least two singular points. If furthermore $Y$ contains at least 3 singular points then either*

(1) *$K_S + X$ is ample and $K_S + 1/2(X + Y)$ is nef, or*

(2) *$S = S(A_1 + A_2 + A_5)$, $X \subset S^0$ is a genus one curve with a simple node, and $K_S + Y$ is numerically trivial.*

*Proof.* If $X$ is smooth, then (13.1) applies. In particular $-(K_S + X + Y)$ is ample and $S$ is smooth away from $X + Y$.

Now suppose $X$ is singular. If there is at most one singular point along $Y$, then $K_S+X+Y$ is negative by adjunction, a contradiction. If $\tilde{Y}$ moves then by (5.11), $\kappa(K_{\tilde{S}} + \tilde{X}) = -\infty$. However since $h^1(K_{\tilde{S}}) = h^0(K_{\tilde{S}}) = 0$, $h^0(K_{\tilde{S}} + \tilde{X}) = h^0(K_{\tilde{X}}) > 0$. Thus $Y$ is a $-1$-curve.

Suppose $K_S + X$ is ample and $K_S + 1/2(X + Y)$ is negative. By (12.1.4) $K_S + X = D$ for some effective $D$ containing all the singularities of $S$ along $X$, and having a non-empty moving part if $g(X) \geq 2$. It follows that the support of $D$ is not contained in $Y$. $K_S+Y+D$ is negative, so by adjunction, $Y$ can contain at most two singular points.

Finally, if $K_S + X$ is numerically trivial, then by adjunction (since $X$ is singular), $X \subset S^0$, $X$ has genus 1, and $S$ is Gorenstein. If $X$ has a cusp, then $S^0$ is simply connected by (7.4.1), and so from the list (3.2), $Y$ contains exactly two singular points. Otherwise $K_S + X$ is log canonical, and by (9.2), $\sum_{p \in S} \frac{r_p - 1}{r_p} \leq 2$.

Suppose $Y$ contains at least three singular points. It follows (since from the list, (3.1), $S$ cannot contain 4 index two points) that $Y$ contains exactly three singularities, and $S$ is smooth off $Y$. Note

$$K_S^2 = \frac{1}{Y^2} = \frac{1}{-1 + \sum \frac{r_p - 1}{r_p}}.$$

It follows (since $K_S^2$ is an integer) that $\sum \frac{r_p-1}{r_p} = 2$, $K_S^2 = 1$, and $\sum r_p = 11$. Thus (from the list) $S = S(A_1 + A_2 + A_5)$. □

**Fences with one smooth side, and one side of genus 1.**

**Proposition 13.4.** *Let $K_S + X + Y$ be a fence, with $Y$ smooth and $g(X) = 1$. Suppose $X$ has an ordinary cusp, and $K_S + 1/2X$ is negative. Further assume either that $X$ contains at most one singular point, or $K_S + 2/3X$ is negative.*

*Then $X \subset S^0$, $S = S(A_1 + A_2)$, and the configuration is obtained from $\mathbb{P}^2$ by taking a flex line to a cubic, blowing up to separate the line and the cubic, then blowing down all the $-2$-curves. In particular $X^2 = 6$, $Y$ contains two singularities, an $A_1$ and an $A_2$, and there are no other singularities on $S$. Moreover, if $\Delta$ is any effective divisor supported on $X + Y$, and $K_S + \Delta$ is anti-nef, then $K_S + \Delta$ has a tiger.*

*Proof.* We begin by showing that $X \subset S^0$. Suppose not. We consider the transformation $S \dashrightarrow W$ of (12.4) (this is where we use the assumption that $K_S + 1/2X$ is negative).

**Claim.** *$W$ has rank one.*

*Proof of Claim.* Note that by (9.2), either $X$ contains a single singular point of $S$, or exactly two $A_1$ points and $S$ is smooth off $X + Y$. Now suppose the claim fails. Then by (12.4.1), $M$ is irreducible, and $X$ contains two $A_1$ points. Since $Y$ is a $-1$-curve, $S$ is not Du Val along $Y$. Thus by (12.1.4), $M$ contains at least three singular points of $S$. This contradicts (12.5). □

Let $G$ be an exceptional divisor adjacent to $X$. $W$ is a rank one Gorenstein del Pezzo, $X \subset W^0$, and $K_W \cdot G = K_W \cdot Y = -1$. $W^0$ is simply connected by (7.4.1). Thus $W = S(E_8)$ by (3.9), and $X$, $G$, $Y$ are all members of $|-K_W|$. But any two members of $|-K_{S(E_8)}|$ intersect in a single point, the unique basepoint of the linear system, by (3.6). But clearly $G \cap X$ and $X \cap Y$ are different points of $W$. Thus we have a contradiction.

So we may assume $X \subset S^0$. $S^0$ is simply connected by (7.4.1). By (13.3) there are two chain singularities on $Y$, so from the list $S = S(A_1 + A_2)$. Blowing up the $A_1$ point and blowing down $Y$ gives the described configuration in $\mathbb{P}^2$.

For the final statement, let $\Delta = aX + bY$. If $a \geq 5/6$, then there is a tiger over the singular point of $X$ (the $-1$-curve of (11.2) configuration III), else $Y$ itself is a tiger, as $K_X + (5/6)X + Y$ is numerically trivial. □

**Proposition 13.5.** *Let $K_S + X + Y$ be a fence, with $Y$ smooth, $g(X) = 1$ and $X \subset S^0$. Assume $X$ has an ordinary node. Then one of the following holds:*

(1) *$S = S(A_1 + A_2)$ with the construction as in (13.4). This is the only possibility with $S^0$ simply connected.*

(2) *$S = S(A_1 + A_5)$. The configuration is obtained from a nodal cubic $X$ and a smooth conic $C$ meeting $X$ to order 6 at a point $p$. Blow up 6 times at $p$, along $C$. $Y$ is the final exceptional divisor over $p$.*

(3) *$S = S(3A_2)$. Begin with a nodal cubic $X$ and two flex lines $L$ and $M$, meeting $A$ at $l$ and $m$. Blow up 3 each times at $l$ along $L$, and $m$ along $M$. $Y$ is the final exceptional*

*divisor over $l$. The singularities along $Y$ are $A_2$ and $(\underline{2}^L, 2^M)$. There is additionally one $A_2$ point away from $X+Y$.*

(4) $S = S(A_2 + A_5)$.

(5) $S = S(A_1 + 2A_3)$, $2A_3 \in Y$.

(6) $S = S(2A_1 + A_3)$, $A_1 + A_3 \in Y$.

(7) $S = S(A_1 + A_2 + A_5)$, $A_1 + A_2 + A_5 \in Y$.

*Proof.* Assume we do not have (7). Then by (13.3) there are exactly two singularities along $Y$, say $A_s$ and $A_t$. $S$ is Du Val. We have $1 < K_S^2 = \frac{1}{Y^2} = \frac{(t+1)(s+1)}{st-1}$. Now one goes through the list (3.1) to check these are the given possibilities for $s$, $t$. It remains now to show that the configurations are as described. For (1-2) one blows up the $A_1$ point and contracts $Y$ just as in the proof of (13.4).

For (3): Suppose first $S$ is any $S(3A_2)$. If $f : T \longrightarrow S$ extracts a $-2$-curve, $E$, then by (3.3), (3.4), and the list, the $K_T$-negative ray gives a birational contraction $\pi : T \longrightarrow S(A_1 + A_2)$. Let $\Sigma$ be the $-1$-curve contracted by $\pi$. It follows that $K_T + \Sigma$ is log terminal, $\Sigma$ contains two $A_2$ points, and meets $E$.

Now apply the above with $E$ one of the $-2$-curves in the $A_2$ point off of $Y$. Since $K^2 = 3$, $K + Y + \Sigma$ is anti-ample. It follows that $Y \cap \Sigma$ is a single point, one of the $A_2$ points along $Y$. Suppose $Y$ and $\Sigma$ meet the same $-2$-curve. Then when we extract this curve, both $Y$ and $\Sigma$ are contractible, a contradiction. Thus $K_S + Y + \Sigma$ is log canonical. Resolve the $A_2$ point $Y \cap W$, and contract $Y$ and $W$ (which are disjoint on the resolution). This gives the described configuration. $\square$

*Remark.* One can also give explicit constructions in (4-6), along the lines of (2-3). We have not done so as we will not need it.

It will be convenient to have a version of the final statement of (13.4) for a slightly more general situation:

**13.6.** *Let $X, Y \subset S$ be two integral rational curves such that $X \subset S^0$ has arithmetic genus one, and an ordinary cusp, and $Y$ meets $X$ at only one point, and normally (but unlike in (13.0), we do not assume $K_S + Y$ is lt at singular points). Then $K_S + \Delta$ has a tiger for any effective $\Delta$ with support contained in $X + Y$ such that $K_S + \Delta$ is anti-nef.*

*Proof.* $X \in |-K_S|$, so $S$ is Gorenstein. $S^0$ is simply connected by (7.4.1). $K_S \cdot Y = -1 = -X \cdot Y$.

Observe that $Y$ is the unique $-1$-curve of (3.8). Indeed if $K_S^2 \geq 2$ then $K_S + Y$ is anti-ample, so $Y$ is smooth and thus a $-1$-curve. Otherwise, by the list, $S = S(E_8)$, $Y \in |-K_S|$, and since $X$ has a cusp, $Y$ is the unique $-1$-curve on $S$, by (3.6). We note that in this case, $Y$ itself is not smooth.

Now we consider the possibilities in (3.8). In (3.8.1-2), $K_S + Y$ is log terminal, so $X + Y$ is a fence and (13.4) applies.

In the remaining cases, (3.8.2-5), $Y$ meets a unique curve, $V$ of the minimal desingularisation. Let $f : T \longrightarrow S$ extract $V$. $Y \subset T^0$ is a $-1$-curve. Let $\pi : T \longrightarrow S_1$ contract $Y$. Let $Y_1 = \pi(V)$. Observe $(S_1, Y_1, X)$ again satisfies the assumptions of (13.6). $K^2_{S_1} = K^2_S - 1$. Scaling as in (8.2.4), we can find $\Delta_1$ supported in $X + Y_1 \subset S_1$, so that a tiger for $K_{S_1} + \Delta_1$ implies a tiger for $K_S + \Delta$. Thus we may induct on $K^2_S$. □

**13.7 Lemma.** *Suppose $C \subset S^0$ is a smooth rational curve. Then either $S = \mathbb{P}^2$, or $S = \overline{\mathbb{F}}_n$ and $C \in |\sigma_n|$.*

*Proof.* If $S$ is smooth, $S = \mathbb{P}^2$. Otherwise let $f : T \longrightarrow S$ extract any exceptional divisor of the minimal resolution. Since $K_S + X$ is negative, and $X \subset T^0$, $T$ must be a smooth net, and $X$ a section. □

## §14 $T_1$ A NET

Here we prove

**14.1 Proposition.** *If $T_1$ is a net, and $\pi_1^{\text{alg}}(S_0^0) = \{1\}$, then $S_0$ has a tiger.*

**We assume throughout §14 that $S_0$ does not have a tiger, and $\pi_1^{\text{alg}}(S_0^0) = \{1\}$.** We will derive a contradiction.

We follow the notation of (8.2.10). Since there are no further hunt steps we drop the subscripts, writing for example $T, S, E, e$ for $T_1, S_0, E_1, e_0$.

We have the $\mathbb{P}^1$-fibration $\pi : T \longrightarrow \mathbb{P}^1$. Recall in particular that $a$ is the coefficient of $E$ in $\Gamma'$, that is $K_T + aE$ is $\pi$-trivial.

$K_T + aE$ is flush, klt, and anti-nef by (8.4.5).

**14.2 Lemma.** *$\pi_1^{\text{alg}}(T^0) = \{1\}$. $E$ is not fibral. Let $d$ be its degree. $d \geq 3$. $e < a = 2/d$. Any singularity along $E$ has spectral value at most one.*

*Proof.* The first claim follows from (7.3). $E^2 < 0$, so $E$ cannot be fibral.

Let $F$ be a smooth fibre. Note $F$ and $E$ generate the two edges of the cone of curve $\overline{NE}_1(T)$.

$$(K_T + E) \cdot E < (K_T + eE) \cdot E = f^*(K_S) \cdot E = 0$$

since $E^2 < 0$.

$$(K_T + E) \cdot F = -2 + d$$

where $d$ is the degree of $E$. $-(K_T + E)$ is not nef (otherwise $E$ is a tiger), so $d \geq 3$.

$$0 = (K_T + aE) \cdot F = -2 + a \cdot d$$

Thus $a = 2/d \leq 2/3$. Now by (8.0.7.2) any singularity along $E$ has spectral value at most 1. □

We get some control from the triviality of the algebraic fundamental group:

**14.3 Lemma.** *If $f : T \longrightarrow \mathbb{P}^1$ is a $\mathbb{P}^1$-fibration with irreducible fibres and with three or more multiple fibres, or two multiple fibres of non-relatively prime multiplicity, then $\pi_1^{\text{alg}}(T^0) \neq \{1\}$.*

*Proof.* If two fibres have non-relatively prime multiplicity, then there is a non-trivial torsion Weil divisor, and hence a finite étale cover of $T^0$. If there are at least three multiple fibres, then there is a branched cover $C \longrightarrow \mathbb{P}^1$, with ramification matching the multiplicities of the fibres (see (6.5) of [23]). The normalisation of the pullback of $T$ gives a cover, étale in codimension one (see (III.9.1) of [4]). □

So by (14.3), $\pi$ has at most two multiple fibres, and if there are two multiple fibres, their multiplicities are relatively prime.

**14.4 Lemma.** $e \geq 1/2$.

*Proof.* Assume $e < 1/2$. By the final remark of (10.8), $S$ has at least three non Du Val singularities, and $1/3 < e < 1/2$. Thus by (10.1), $x = (3, A_r)$ for $r > 1$ and $E$ contains a unique singularity, $y$, an $A_r$ point. $e = \frac{r+1}{2r+3} \geq 2/5$. Thus by (14.2), $d \leq 4$. Let $G_1$ be the multiple fibre containing $y$. By (11.5.5), there is exactly one other multiple fibre, $G_2$, and it contains a non Du Val point.

The possible non Du Val fibres, by (11.5.9), are (11.5.9.2), (11.5.9.6), and (11.5.9.5) with $k = 2$, which have respective multiplicities three, five, and three.

(11.5.9.6) is not possible, for $E$ must meet such a fibre at a singular point (otherwise $d$ is at least 5), a contradiction, as this fibre contains no Du Val singularity (and the only singularity on $E$ is Du Val).

Thus any non Du Val fibre has multiplicity three. In particular this holds for $G_2$. Thus by (14.3), $G_1$ is Du Val. $G_2$ then must contain two non Du Val points, contradicting the descriptions (11.5.9.2) and (11.5.9.5). □

By (14.2) and (14.4), $d = 3$.

We let $h : \tilde{T} \longrightarrow W$ be a relative minimal model, as in (11.5.4). By (11.5.5) there are two choices for $h$ at each multiple fibre.

**14.5 Lemma.** *Let $G$ be a multiple fibre of $\pi$. We may choose $h$ so that $h(E)$ is smooth at the image fibre, and $\tilde{E}^2$ goes up by at most 4. Further if $\tilde{E}^2$ changes by 4, then $G$ is uniquely determined.*

*Proof.* We use the classification (11.5.9). We check first that it applies. By (14.2), the singularities along $E$ have spectral value at most 1, so if $G$ meets $E$ in a singular point, then by (8.0.8.b), (11.5.9) applies. Otherwise $G$ meets $E$ in a single smooth point, and so the multiplicity of $G$ is three, and we can apply (11.5.13).

We will indicate the type of the fibre (using the notation of (11.5.4)), the change in $\tilde{E}^2$ under $h$, and leave the choice of $h$ to the reader. We let $c$ be the change in $\tilde{E}^2$ (for optimal choice of $h$). If $G$ meets $E$ in two singular points, then the fibre is

$$(-\underline{3}) + (-1) + (-\underline{2}) + (-2), \quad c = 2.$$

If $G$ meets $E$ in one singular point, and one smooth point, then the fibre is

$$(-\underline{2}) + (-\underline{1}) + (-2), \quad c = 2,$$

or given by (11.5.9.1) and $c = 3$. If $G$ meets $E$ in one point, a singular point, and $\tilde{E}$ and $\tilde{G}$ meet, then the fibre is again $(-2) + (-1) + (-2)$, $c = 2$. If $G$ and $E$ meet only once, at a smooth point, then $m = 3$, and the fibre is given by (11.5.9.2) or (11.5.9.5), with $c = 4$, or 3 respectively. Finally if $\tilde{E}$ and $\tilde{G}$ are disjoint, and $G$ meets $E$ in a single singular point, then the fibre is given by (11.5.9.5) with $k = 3$, or (11.5.9.6), and in either case $c = 3$. □

*Proof of (14.1).* As remarked below (14.3), $\pi$ has at most two multiple fibres, and cannot have two of the same multiplicity. Thus by (14.5) we may choose $h$ so that $h(E)$ is a smooth triple section of self-intersection at most 5 (corresponding to the case when $\tilde{E}^2 = -2$, there are two multiple fibres, and, under $h$, $\tilde{E}^2$ goes up by three at one multiple fibre, and four at the other). But by (11.5.1)0, $h(E)^2 = 2d = 6$. Thus we have a contradiction. □

## §15 $g(A) > 1$.

Here we prove

**15.0 Proposition.** *If $S = S_0$ does not have a tiger and $g(A_1) \geq 2$, then $S$ is the surface in (15.2), and $S^0$ is uniruled.*

**Throughout §15 we assume $S = S_0$ has no tiger.**

$1/3 \leq e_0 < a$, by (10.8), and so by (8.3.7) $A$ either has multiplicity three, or a double point. By (8.4.7), $K_{T_1} + a_1 E_1$ is flush. The possible configurations for $T_1 \longrightarrow S_1$ (in a neighbourhood of $\Sigma_1$) are classified in (11.1-3).

We will repeatedly use the hunt notation (8.2.10), the notation of (11.0.1) to describe the contraction $\pi : T_1 \longrightarrow S_1$, and the notation of (10.2).

We begin by ruling out multiplicity three:

**15.1 Lemma.** *$A_1$ cannot have multiplicity three.*

*Proof.* Suppose $A_1$ has multiplicity three. The possible local pictures for $\pi$ are given in (11.3), with $W = E_1$. By (8.0.9.1), $e_0 < 1/2$, so $x_0 = (3, A_s)$ for some $s$, by (10.1). By (10.8), $K^2_{\bar{S}_0} = -1$.

Consider what happens when $\pi$ is given by (11.3.2) or (11.3.3.2) (numerically these cases are indistinguishable). In the notation of (11.0.1), $X$, $Y$ are the two analytic discs where $E$ meets $\Sigma$. From the description of the configuration, $x_0 = (3, 2)$. By (10.8) there is a non Du Val point away from $\overline{\Sigma}$. In particular $\alpha \geq 1/3$ ($\alpha$ of (10.2)). By (10.9)

$$K^2_{S_0} = -1 + 1/3 + 2/5 + \alpha = \frac{(K_{S_0} \cdot \Sigma)^2}{\Sigma^2} = \frac{(-1 + 1/3 + 3/5)^2}{(-1 + 1/3 + 7/5)}.$$

and so $\alpha = 3/11 < 1/3$, a contradiction.

Now consider (11.3.3.1). Since there is at least a $(2, 3)$ singularity, $x_0 = (3, A_s)$ with $s \geq 1$. By (10.6) there are no singularities away from $\overline{\Sigma}$. As $\rho = 10$, $s = 6$. But then the singularities along $\overline{\Sigma}$ are exactly $(3')$, $(2', 3)$ and $(3', A_6)$ and so $K_{S_0} \cdot \overline{\Sigma} = 0$, a contradiction. □

We now turn to double points of genus $g \geq 2$.

We consider in turn each possible configuration on $T_1$ from the classification (11.1-2). We note that configuration 0 for a node cannot occur (since $E$ is smooth). In the case of the node we apply the classification with $X$ and $Y$ (in the notation of (11.0.1)) the two branches of $A$ at $q$ (that is the two branches of $E$ along $\Sigma$), and $c = d = a$. For the cusp, we take $X = A$, $c = a$. We note that the node and cusp configurations labeled by the same roman numeral are numerically indistinguishable.

**Configuration $I$.** In this subsection we also allow $g = 1$, since most of the analysis in this case is the same (and this will save a little work in §16-17).

By (11.1-2.1), $a = 1/2$. Thus $1/3 \leq e_0 < 1/2$. By (10.8), $K^2_{\bar{S}_0} = -1$, $e_0 > 1/3$. By (10.1), $x_0 = (3, A_r)$ with $r \geq 1$, and $e_0 = \frac{r+1}{2r+3} \geq 2/5$. By the description of Configuration I, $A_1 \subset S_1$ contains a unique singular point, an $A_r$ point. By (10.9) we have

$$-1 + e_0 + \alpha = \frac{(K_{S_0} \cdot \overline{\Sigma})^2}{\overline{\Sigma}^2} = \frac{(-1 + 2e_0)^2}{4e_0 + \frac{g-1}{g} - 1}$$

so that $\alpha = \frac{e_0 + g - 1}{4e_0 g - 1}$. By (10.8), $n \geq 3$, so there are at least two non Du Val singularities off $\overline{\Sigma}$. Thus $\alpha \geq 2/3$. Since $e_0 = \frac{r+1}{2r+3} \geq 2/5$, $\alpha \leq \frac{5g-3}{8g-5} \leq 2/3$. Thus we must have equality in each of the inequalities, $r = g = 1$ and the non Du Val singularities of $S_0$ are exactly $(2, 3), (3), (3)$. By (10.6.2) there is exactly one other singular point, an $A_6$ point. Now consider the transformation $S_1 \dashrightarrow W$ of (12.4). $W$ has rank one by (12.4.1). The $-1$-curve $M$ (of (12.0)) is irreducible,

$K_{S_1} + M$ is negative, and $M$ contains the singular points $(3), (3), (2)$ of $S_1$, by (12.1.3-4). By adjunction $M$ cannot contain any other singular points, and $K_{S_1} + \overline{M}$ is log terminal, where $\overline{M}$ is the reduction of $M$. Thus $W = S(A_2 + A_6)$. But there is no such surface in the list (3.1). □

**Configuration** $II$.

By (11.1.1) and (11.2.1) $e_0 < a = \frac{g+1}{2g+1} \leq 3/5$, and so by (10.1) either $x_0 = (2,3,2,2)$ and $g = 2$, or $x_0 = (A_g, 3)$.

Suppose $x_0 = (3, A_g)$. $e_0 < 1/2$, so the final remark of (10.8) holds. Let $V$ be the $-2$-curve which meets $\Sigma$. One checks $\overline{\Sigma}$ has coefficient $\frac{4g}{2g+3}$ on $V$ and $\frac{2g+1}{2g+3}$ on $E$. Also $e_0 = \frac{g+1}{2g+3}$. By (10.9) we have

$$-1 + \frac{g+1}{2g+3} + \alpha = \frac{(-1 + \frac{2g+1}{2g+3})^2}{\frac{6g+1}{2g+3} - 1}$$

so that $\alpha = \frac{g}{2g-1}$. Since $n \geq 3$, $\alpha \geq 2/3$, so it follows that $g = 2$ and $\alpha = 2/3$. Thus there are exactly two other non-Du Val points, each (3). Since $\rho = 10$ there is another singular point, $z$, the only other singularity, by (9.2). Thus $\rho(z) = 5$. But then $z$ has index at least 6, and we have singularities of index at least $(3,3,6,7)$, which violates (9.2).

Now suppose $x_0 = (2,2,3,2)$. By (10.11), $E$ is the $-3$-curve. Then, by the description of configuration II, $A$ contains a unique singular point, $w$, of $S_1$, an $A_1$ point. $M$ (of §12) is irreducible or empty by (12.1.7). $K_{S_1} + M + F$ is negative, and so by (12.1.4) and adjunction, $n < 4$. $e_0 = 6/11$, so we have one of (10.8.3-5). In particular, $K^2_{\bar{S}_0} = -1$ or $-2$. We have

$$K^2_{\bar{S}_0} + 6/11 + \alpha = \frac{(-1 + 6/11 + 4/11)^2}{-1 + 24/11}$$

thus $\alpha = 6/13 < 1/2$ or $\alpha = 1 + 6/13$, depending on whether $K^2_{\bar{S}_0}$ is $-1$ or $-2$.

Suppose we have the second possibility. Then by (10.8), $n = 3$, $w = 4$, thus there are two more singular points, one, $y$ of weight one, and the other, $z$, of weight two. The only weight two singularities of coefficient at most $e_0 = 6/11$ are given by (10.1.2). In particular, $\delta(z) = 1$. Hence $\delta(y) = e(y) = 6/13$. By (10.1), $y = (3, A_5)$. Since $\rho = 11$, $\rho(z) = 1$, $z = (4)$, and there are no other singular points. By (12.3.5) $M \subset S_1$ is nonempty, and must pass through (2) and (4). By adjunction $M$ is smooth. But then $M$ contracts on $S_1$, a contradiction.

Thus we have the first possibility. Necessarily there is exactly one more non Du Val singularity, $y$, and $w(y) = 1$. So

$$\alpha = \delta(y) = e(y) = 6/13$$

and thus $y = (3, A_5)$ by (10.1). Since $\rho = 10$, there are no other singular points. We now show there is exactly one possibility for $S = S_0$.

**Notation:** *For a conic $X \subset \mathbb{P}^2$ and points $w, z, p \in X$, we let $L_{wz}$ and $M_p$ indicate the secant through $w, z$, and the tangent line through $p$.*

**15.2 Definition-Lemma.** *Let $A$ and $B$ be two smooth conics meeting to order four in $\mathbb{P}^2$ at a point $c$. Pick a point $b \neq c$ of $B$. Pick a point $a$ of $A \cap M_b$. Let $b'$ be the other point of $L_{ac} \cap B$. Let $a'$ be a point of $A \cap L_{bb'}$.*

*Let $f : \tilde{S} \longrightarrow \mathbb{P}^2$ blow up once at $a$, $a'$, $b'$, twice along $M_b$ at $b$, and five times along $A$ at $c$. We have*

(1) *$\tilde{S}$ is the minimal resolution of $S$.*
(2) *All configurations $(A, B, b)$ are projectively isomorphic. In particular $S$ is unique.*
(3) *$L_{aa'}$ meets $B \setminus \{b, b', c\}$ in two distinct points.*
(4) *$L_{bb'}$ is not tangent to $A$. Equivalently $q$ is not a unibranch singularity of $A_1$.*
(5) *$S^0$ is uniruled.*

*Proof.* Let $Z \longrightarrow S_1$ extract the divisors adjacent to $F$ (of §12). Then $F$ is the general fibre of a $\mathbb{P}^1$-fibration on $Z$, by (12.3).

We argue first that $y, w \in F$. Otherwise $Z$ has Picard number two, and $Z \longrightarrow S_1$ extracts a unique divisor $E$, a section of the fibration. $E$ cannot be over $w = A_1$, for otherwise $E \subset Z^0$, and so (since $E$ is a section) $E$ is smooth, a contradiction, as $y \in Z$. Then by (12.3.3), $E$ is the $-3$-curve of $y$, so $Z$ is Du Val, and has an $A_5$ singularity, contradicting (11.5.4).

Now we claim $M$ is empty: As we have observed, it has at most one irreducible component. Since $y$, $w \in F$, and $K_{S_1} + F + M$ is negative, $M$ can only contain one singular point by adjunction, and so by (12.1.6), $w \in M$. But then $M$ contracts on $S_1$, a contradiction. So $M$ is empty. Then by (12.3), $E$ is empty as well, and $K + \tilde{A} = \tilde{F}$.

$Z \longrightarrow S_1$ extracts (by (123.3)) the divisors $M_b$, the $-2$-curve of $w$, and $B$, the $-3$-curve of $y$. $Z$ has Picard number three. There is a unique reducible fibre $F_1 + F_2$, and $A \subset Z^0$. $Z$ is Du Val, with unique singular point $v = A_5 \in B$. $F_1$ and $F_2$ are contractible $-1$-curves, and so the singularities along either are described by (3.3). $K_Z + A = F$, so $A$ meets each $F_i$ normally in a single (smooth) point. $v$ cannot lie on an irreducible fibre, by (3.4). Now $F_i$ is not in the smooth locus, or after contracting it, $v$ lies on an irreducible fibre. It follows that $v = F_1 \cap F_2$ and $K_Z + F_1 + F_2$ is lc at $v$.

Let $F_1$ be the component meeting $M_b$ (there can be only one as $M_b$ is a section contained in $Z^0$). If we contract $F_2$ we obtain $\mathbb{F}_2$, $M_b$ becomes $\sigma_\infty$ and $B$, as it remains disjoint from $M_b$, becomes $\sigma_2$. It follows that $K_Z + B + F_1$ is lc at $v$, and $F_2$ and $B$ meet the same curve over $v$.

Let $t \in \mathbb{F}_2$ be the image of $F_2$. Then $A$ and $B$ meet only at $t$, with fifth order contact.

Consider extracting $V$, the $-1$-curve of the second blow up at $q$ along $A$, and blowing down the fibre (through $q$). The resulting surface is $\mathbb{P}^1 \times \mathbb{P}^1$, where $M_b$ becomes a zero curve, and $A$ a section (in the direction for which $M_b$ is a fibre). Let $L$ be the unique curve disjoint from $M_b$ passing through $t$. $L$ is a 2-curve on $\mathbb{F}_2$ through $t$ and $q$, and disjoint from $M_b$. $A$ meets $L$ to order four at $q$. $A$ and $L$ are transverse at $t$ (as is clear on $\mathbb{P}^1 \times \mathbb{P}^1$) thus $B$ meets $L$ transversally at $t$ as well.

Let $L_{bb'}$ be the marked curve $(2, 2^{L_{bb'}}, 3^A, 2)$ over $x_0$ and $L_{ac}$ the marked curve $(3^B, A_4, 2^{L_{ac}})$ at $y$. Let $G$ be the fibre of the ruling (on $Z$) through $q$.

Then $f$ is given by contracting $\Sigma$, $G$, $F_1$, $F_2$, $L$ and all exceptional curves over $S$, other than $A$, $B$, $L_{bb'}$, $L_{ac}$, $M_b$. Hence (1).

Pairs $(C, p)$ of a point on a smooth conic are projectively equivalent, so we may assume $A$ is the conic $yz = x^2$ and $c$ is the origin $(0, 0, 1)$. Then $B$ is in the pencil of conics with $4^{th}$ order contact at $c$, $yz - x^2 + ty^2 = 0$, for $t \neq 0, \infty$. The automorphisms $(x, y, z) \longrightarrow (\lambda x, \lambda^2 y, z)$ (for $\lambda \neq 0$) act transitively on the pencil elements with $t \neq 0, \infty$. Thus we may assume $B : yz - x^2 + y^2 = 0$. The automorphisms

$$\phi_d : (x, y, z) \longrightarrow (x + \frac{d}{2} \cdot y, y, d \cdot x + \frac{d^2}{4} \cdot y + z)$$

for $d \in \mathbb{A}^1$ preserve $A$ and $B$. Acting on a general point of $B$ gives a non-constant map $\mathbb{A}^1 \longrightarrow B \setminus \{c\} = \mathbb{A}^1$, which is necessarily a surjection. Hence (2).

By (2) we can assume $b$ is the point $(0, -1, 1)$. Now (3) and (4) are straightforward calculations, which we leave to the reader.

By (3) and (4), $\overline{\Sigma}$ and $L_{aa'}$ are each $-1$-curves on $S$, meeting $\mathrm{Sing}(S)$ twice. $\overline{\Sigma}$ has branches $(2, 2', 3, 2)$ and $(2, 2, 3', 2)$, both at $x$, while $L_{aa'}$ has two branches, each $(A_5, 3')$ at $y$. Every branches is smooth by (4.12.2). The indices of $x$ and $y$ are 11 and 13. Thus $S^0$ is uniruled by (4.10.3) and (6.6). □

*15.2.6 Remark.* It is tempting to think that (15.2.2) follows immediately from the fact that $A + B$ is canonically embedded. However, the scheme structure on $A + B$ is not so obvious. If one constructs a scheme $Z(\phi)$ by gluing together two copies of $\mathbb{P}^1$ along an automorphism $\phi$ of the length 4 curvilinear scheme supported at some point, then the isomorphism class of $Z(\phi)$ depends in general on $\phi$. For example if one takes the identity for $\phi$, then the canonical map for the resulting Gorenstein curve (of arithmetic genus 3) is the two to one map to $\mathbb{P}^1$ which is the identity on each component. (15.2.2) can be viewed as saying that any two $Z(\phi)$ for which the canonical map is an embedding, are isomorphic.

Note by (11.1-2) that for further configurations, $q \in A$ is a unibranch singularity. The possibilities are classified in (11.2.1.3-5). We consider each in turn.

**Configuration *III*.**

Let $y$ be the $(A_{g-1}, 3)$ point along $\Sigma$. We have by (10.1.1) and (11.2.1.3)

$$e(y) = \frac{g}{2g+1} \leq e_0 < a = \frac{g+1}{2g+1} \leq 3/5.$$

Thus

$$\frac{2g}{2g+1} \leq e_0 + e_y < 1.$$

One easily rules out (10.8.1), (10.8.2) and (10.8.5). So we may assume there is exactly one other non Du Val singularity, $z$, of $S_0$, of weight one or two, and $w \leq 4$.

Suppose $z$ has weight two. Then we have (10.8.3). As $\beta > 1/2$, $1/2 < e(z) < 3/5$. Thus by (10.1), $z = (4, 2)$. $e(z) = 4/7 < a = \frac{g+1}{2g+1}$, thus $g = 2$. By (9.2) there are no other singularities. Since $\rho = 11$, $\rho(x) = 6$. $x$ has weight one, and $1/2 < e(z) < 3/5$, so by (10.1), $x = (2, 3, A_4)$ which has index 17. But now we have singularities with indices $(2, 5, 7, 17)$, which contradicts (9.2).

Thus $z$ has weight one. Suppose first $x$ has weight two. From $1/2 < e(z) < 3/5$, it follows from (10.1) that $x = (4, 2)$. Exactly as in the previous paragraph, $g = 2$ there are no other singularities, and $\rho(z) = 6$. But then $z$ has index at least 13, and we have indices of at least $(2, 5, 7, 13)$, which contradicts (9.2).

Thus $x$ has weight one as well, and we have (10.8.4). $K^2_{\tilde{S}_0} = -1$, and $\rho = 10$. By (10.9) we have

$$-1 + e_0 + e_y + \alpha = \frac{(-1 + e_y + e_0)^2}{-1 + e_y + e_0 + 1/2}$$

so that $\alpha = \frac{1 - e_0 - e_y}{2e_0 + 2e_y - 1}$. As $\alpha \geq 1/3$, it follows that $e_0 + e_y \leq 4/5$. Thus $\alpha = 1/3$, $e_0 = e_y = 2/5$, and thus the non Du Val singularities are exactly

$$x, y, z = (2, 3), (2, 3), (3).$$

By (9.2) there is at most one other singularity, and any additional singularity can have index at most 3. But then $\rho \leq 7$, a contradiction. □

**Configurations $u$ or $v$.**

The final possibilities are the genus two unibranch singularities given by (11.2.1.4) and (11.2.1.5). In the first case, configuration $u$, there are two singularities along $\Sigma \subset T$, $(2, 4')$ and $(\underline{2}', 2)$. In the second case, configuration $v$, there is one singularity along $\Sigma$, $(2, 3, 2', 2)$, and $E$ meets $\Sigma$ normally at a smooth point.

**15.3 Lemma.** *If we have configuration u or v then $S_1$ has at least two singular points along $F$ ($F$ of (12.3)).*

*Proof.* $\tilde{F} \subset \tilde{S}$ is a fibre, so obviously $F \not\subset S_1^0$. Suppose there is a unique singular point, $b$ of $S_1$ along $F$.

Let $f : Z \longrightarrow \mathbb{P}^1$ be the $\mathbb{P}^1$-fibration of (12.3) obtained by extracting the unique divisor $E$, adjacent to $F$. Note $E$ is a section, and by (12.3.1) $A$ is a double section. As $\rho(S_0) \geq 10$, $\rho(Z) \geq 4$, and so there is at least one multiple fibre $F'$.

Let $\Gamma$ be the fibre passing through the singularity $q$ of $A$, $\Gamma_q$ the local analytic branch of $\Gamma$ at $q$, and $\Gamma_e$ the local branch where $\Gamma$ meets $E$.

Then $\Gamma$ is a smooth fibre and $q$ is a ramification point of $f|_A$.

Since $A \cdot \Gamma = 2$, $\Gamma_q$ and $A$ are separated by the first blow up of $\tilde{T} \longrightarrow S_1$ at $q$. $x_0 \notin \Gamma_q$ and by (11.2.1) the singularity of $\Gamma_q$ is respectively $(2', 4)$ for $u$, and $(2', 3, 2, 2)$ for $v$. Thus $K_{S_0} + \Gamma_q$ is lt. $\Gamma_e$ and $\Gamma_q$ pass through different points of $S_0$, and are the only branches of $\Gamma$ through singular points of $S_0$. Thus $K_{S_0} + \Gamma_e$ is not lc, for otherwise $\Gamma$ is a tiger.

Suppose $b \notin A$. Thus $K_{S_1} + \Gamma_e$ is not lc at $b$. A multiple fibre meets $A$ in only one point, for otherwise it will contain three singularities (the two points where it meets $A$ and the point where it meets $E$), contradicting (11.5.5). As $f|_A$ ramifies at only two points, $G$ is the unique multiple fibre, and there is a unique singular point $b'$ on $E$. Thus $b$ is not a chain singularity (otherwise $K_{S_1} + \Gamma_e$ is lt). By (12.3.3) $E$ is not a $-2$-curve, and as $e_0 < 2/3$, $b$ is given by (10.1.2.d), and $G$ meets the unique $-3$-curve. But then $K_{S_1} + \Gamma_e$ is lc (see Appendix $L$), a contradiction.

Suppose $b \in A$. $x_0$ is a chain singularity by (10.1) since $1/2 < e_0 < 2/3$. Thus $\Gamma_e$ does not meet an end of the chain at $x_0$. As $a < 2/3$, by (8.0.7.2), any singularity along $A$ has spectral value at most 1 (with respect to $A$), and so by (8.0.8) and (12.3.3), $K_{S_1} + F$ is plt at $b$. Thus $E$ is an end of the chain at $b$, and not at $x_0$. Thus $A \cap E$ is a smooth point of $Z$, there is a unique singular point $b'$ on $E$, and $G$ is the unique multiple fibre. As $b'$ is a chain singularity, Du Val or almost Du Val, $G$ is described by (11.5.9). If $m \neq 2$, since $A$ is a double section, by (11.5.5), $G$ meets $A$ at a unique point, $b'' \neq b$, so $G$ contains two chain singularities, each either Du Val or almost Du Val. This rules out cases (2-4), (7-8), and (5) for $k \geq 3$. $A$ meets a curve over $b''$ of multiplicity two in the fibre, thus in (6) the marked singularity $(A, b'')$ is $(3', 2)$, which has spectral value 2, a contradiction. In the remaining cases, which are (1), and (5) with $k \leq 2$, $\rho(Z) \leq 3$, a contradiction. □

**15.4 Lemma.** *If we have type $u$ or $v$ then $S_0$ has a tiger.*

*Proof.* By (15.3) we may suppose $F$ passes through at least two singular points.

We show first that $M$ (of §12) is empty. Suppose not. Then by (12.1.2), $M \cap F$ is a single point $b$. So there is a singularity along $F \setminus M$. Let $r$ be the index of one such singularity.

Let $N$ be an irreducible component of $M$. Note that as $K_S + (7/11)A$ is negative, so is $K_S + (7/4)(M + F)$. Let $V$ be the curve which $F$ meets at $b$, and let $\eta = e(V, K_{S_1})$. Note $e(V, K_{S_1} + N + F) \geq 1$ and so the coefficient of $V$ in $N + F$ is at least $1 - \eta$.

We have

$$0 > (K + (7/4)(N + F)) \cdot F \geq -2 + \frac{r-1}{r} + 1 + 3/4(N + F) \cdot F \geq -1/r + 3/4(1 - \eta + 1/r).$$

It follows that $2/3 > a > \eta > \frac{3r-1}{3r} \geq 5/6$, a contradiction.

Hence $M$ is empty. The singularities along $A$, as well as any non Du Val singularities of $S_1$ away from $A$, are described by (12.1.4) and (12.3.3-4). Let $r_1, \ldots, r_t$ be the indices of the singular points of $A \cap F$ (when there are no singular points, take $t = 1$, $r_1 = 1$). Since $K_{S_1} + A$ is lt at singular points (by (8.4.5) and (8.0.4)), and $A$ is in the Du Val locus

$$A \cdot F = 2 + \sum \frac{r_i - 1}{r_i}.$$

Note that $K \cdot F = -2 + \alpha$ ($\alpha$ as in (10.2)).

By (11.2.1.4-5)

$$-2 + \alpha + 7/11(2 + \sum \frac{r_i - 1}{r_i}) = (K_S + 7/11A) \cdot F < 0. \tag{15.4.1}$$

Suppose first $t \geq 2$. Then (15.4.1) implies $t = 2$, $r_1 = r_2 = 2$ and $\alpha = 0$. Thus $S_1$ is Du Val and (by the configuration on $T$) $n = 2$. Hence by (10.8), $w = 2$. Since $x_0$ has weight at least one, the configuration cannot be $u$ (which has a $-4$-curve away from $x_0$), thus the configuration is $v$, $E$ is a $-3$-curve, and $x_0$ is a non-chain singularity with branches $(2,2)$, $(2)$ and $(2)$. But then $e_0 \geq 3/4$, a contradiction.

Next suppose there are at least two singular points along $F \setminus F \cap A$. Then (15.4.1) and (10.1) imply that there are exactly two singularities on $F$, each of type $(3)$. Thus $n = 4$, and so by (10.8), three of the singularities on $S_0$ are $(3)$, clearly impossible.

Thus we can assume there is exactly one singularity, along $A$ (necessarily in $A \cap F$) and exactly one non Du Val singularity, $b$ (necessarily in $F \setminus F \cap A$) of $S_1$. (15.4.1) implies $r_1 = 2$, and $b = (3)$ or $(3,2)$. $x_0$ cannot have weight one (or there is another singularity of higher coefficient). If the configuration is $u$, then $e_0 \geq e(2,2,4,2) > 2/3$, a contradiction. Thus the configuration is $v$, and $x_0 = (2,4)$. Thus $n = 3$ and $w = 4$, and so by (10.8), $\rho(S_0) = 11$. By (9.2) there is at most one other singular point. Since $\rho(S_1) = \rho(S_0) - 5 = 6$, there must be exactly one more singular point, Du Val, of index at least 4. But now we have indices on $S_0$ of at least $(4,3,7,11)$, which violates (9.2). $\square$

## §16 $A_1$ HAS A SIMPLE CUSP

This section is devoted to proving:

**16.1 Proposition.** *If $A_1$ has a cusp of genus one, and $\pi_1^{\text{alg}}(S^0) = \{1\}$, then $S$ has a tiger.*

**Throughout §16 we assume $S$ has no tiger, and $\pi_1^{\text{alg}}(S^0) = \{1\}$.** Our goal will be to obtain a contradiction.

$K_{T_1} + a\Sigma$ and $K_{S_1} + aA$ are flush by (8.4.7). The possible configurations for $T_1 \longrightarrow S_1$ are given by (11.2). As Case I has been ruled out in (§15), we may assume by (11.2.1) that $a \geq 2/3$.

Note by (7.3.1), since $q = q_1 = \pi_1(\Sigma_1)$ is a smooth point of $S_1$, $\pi_1^{alg}(S_1^0)$ is trivial. $A \not\subset S_1^0$, for otherwise $K_{S_1} + A$ is numerically trivial, and $A$ is a tiger.

By (11.2.1), $2/3 \leq a_1 < 4/5$. By (8.3.8), $e_1 \geq 1/3$, so $a_1 + e_1 \geq 1$. Since $A \neq \Sigma_2$ ($\Sigma_2$ is smooth) it follows from (8.4.7) that:

*16.2.* The possibilities for the next hunt step are:

(1) $A_2 + B_2$ is a tacnode

(2) $A_2 + B_2$ is a banana

(3) $A_2 + B_2$ is a fence

(4) $T_2$ is a net

Recall that $E_1$ is a $-k$-curve, and $E_2$ is a $-j$-curve.

For the next lemma (and its proof) we use the notation of §12.

**16.3 Lemma.** *The following hold:*

(1) *$S_2$ is not a fence,*

(2) *$W$ (of (12.4)) has rank at least two,*

(3) *$S_1$ is singular along $A$ in at least two points,*

(4) *$S_1$ is Du Val outside $A$,*

(5) *$S_1$ is not Gorenstein, and*

(6) *$S_2$ is not a tacnode.*

*Proof.* Suppose $S_2$ is a fence. Then by (8.4.7.8.1), $-(K_{S_2} + B)$ is ample, so $B$ is smooth by adjunction. But then by (13.4), there is a tiger, a contradiction.

Suppose $W$ has rank one.

As $A$ is simply connected, $W$ is also, by (7.4.1).

Claim: There is only one singularity along $A$: Otherwise by (12.4) and (3.9), $W = S(E_8)$, (for some $i$) $D_i \subset W^0$, $D_i \in |-K_W|$. But as $A \in |-K_W|$ has a cusp, this contradicts (3.6).

Now $M$ is irreducible by (12.4.1).

Now define $\Gamma$ by $K_Y + \Gamma = f^*(K_S + aA)$, and define $\Gamma' = \lambda\Gamma$ such that $K_Y + \Gamma'$ is $\pi$ trivial (as in the definition of the hunt), let $\Delta' = \pi(\Gamma')$. $K_W + \Delta'$ has a tiger by (13.6). But then $K_S + aA$ has a tiger, a contradiction.

(3) follows from (2) and (12.4.1).

If there is a non Du Val singularity outside $A$, then $M$ passes through that point by (12.1.4). Say there are $l$ singular points along $A$. Then by (2) and (12.4.1), $M$ has at most $l-1$ components, and contains at least $l+1$ points. By (12.1.2) some component of $M$ contains at least three singular points, contradicting (12.5). Hence (4).

Suppose $S_1$ is Du Val. By (3) and the classification of simply connected Gorenstein surfaces $S_1 = S(A_1 + A_2)$. Thus

$$\tilde{A}_1^2 = K_W^2 > K_Y^2 = K_{S_1}^2 = 6,$$

which contradicts (16.4) below. Hence (5).

Thus $S_1$ has a non Du Val point along $A_1$. (6) follows exactly as in the second paragraph of the proof of (18.5). □

**16.4 Lemma.** *If $\tilde{A}_1^2 \geq 7$, then $A$ has three singularities along it.*

*Proof.* By (11.2.1) the self-intersection of $E$ can only go up by nine (corresponding to $(v, n^2)$). Thus $k = -2$ and $x_0$ is not a chain singularity, by (10.11). □

**16.5 Lemma.** *(16.2.2) does not occur.*

*Proof.* Suppose not.

$B \not\subset S_2^0$ (or $B$ is a tiger), so there is a singularity $y \in B$. $x_1 \in A$ (the only way to get a Banana), so there is a unique singularity along $E_2$, by (8.0.7.1), and $\Sigma$ meets $E$ at a smooth point. Since $a_1 + e_1 \geq 1$, $\Sigma$ must meet $A$ at a singular point $z$. The configuration along $\Sigma_2$ is given by (11.1.1.3). In particular $z$ is a Du Val point.

As there is only one singularity along $B$, $B$ is not a $-1$-curve. On the other hand $-K_S \cdot B \leq A \cdot B = 2$. It follows that $B$ is a 0-curve.

Suppose $S_2$ is singular along $A_2$. Then $y$ is not a Du Val point, $x_0$ is not a chain singularity (as there are at least 3 singularities along $A_1$, and hence along $E_1$). $x_1$ is at least $(2,3)$. Since $e_0 < 4/5$, one checks using (8.3.9) that the only possibility is that $k = j = 2$, and the branches of $x_0$ are $(2)$, $(2)$ and $x_1 = (A_r, 3)$, $r \geq 1$. Then $z = A_1$, $A_2$ contains an $A_1$ point, $y = (A_{r-1}, 3)$, and by (9.2) there are no other singularities. Let $\pi : T \longrightarrow S_2$ extract the $-2$-curve, $E_3$, adjacent to $B$ (this is in fact the next hunt step, but we will not use this). $T$ is a net, $B$ is a smooth fibre, and $E_3$ is a section, containing a unique singularity, $(A_{r-2}, 3)$ (by which we mean $(3)$ if $r \leq 2$). Thus the fibre through the $(A_{r-2}, 3)$ point is the unique multiple fibre, and so this fibre contains the $A_1$ point as well. But there is no such net by (11.5.9).

Thus $A_2$ lies in the smooth locus and $S_2$ is Gorenstein. As $y$ is a chain singularity and $S_2$ is simply connected, the possibilities for $S_2$ are $S(A_1)$, $S(A_1+A_2)$, or $S(A_4)$. Moreover $x_1 = (j, y)$, $z = A_{j-1}$ (since $B$ is a 0-curve) and $\tilde{A}_1^2 = K_{S_2}^2 - 1$. Now $j \geq 3$, by (16.3.5). Since the spectral value is at most three by (8.0.7.2), $y = (2)$ or $(2,2)$ and $j = 3$, by (8.0.8). If $y = (2,2)$, then $e_1 > 3/4$ and $a/3 + b > 1$, so $(K_{T_2} + \Gamma') \cdot \Sigma_2 > 0$, contradiction.

Thus $y = (2)$ and by (16.4), and the simply connected list, $S_2 = S(A_1 + A_2)$. But then if $f : T \longrightarrow S_2$ is the blow up at $y$, $T$ is a Du Val net, with exactly one singularity, an $A_2$ point, contradicting (3.4). □

For the next Lemma, we allow the possibility that $A$ has a node (as the analysis will be the same in that case).

**16.6 Lemma.** *(16.2.4) does not occur.*

*Proof.* Suppose $T_2$ is a net. As $a_1 \geq 2/3$, necessarily $E_2$ is a section and $A$ is a double section. By (11.5.11.2), $E$ contains at most two singularities and so $x_1$ is a chain singularity. In particular $j \geq 3$. But by (11.5.10), $E_2$ is a $-2$-curve, a contradiction. □

As we have eliminated every case of (16.2), this completes the proof of (16.1).

## §17 $A_1$ HAS A SIMPLE NODE

Our goal is to prove:

**17.1 Proposition.** *If $\pi_1^{\text{alg}}(S^0) = \{1\}$ and $g(A_1) = 1$, and $A_1$ has a node, then either $S_0$ has a tiger, or $(S_2, A_2 + B_2)$ is a fence, with $B_2$ smooth. In any case $S_0$ is log uniruled.*

For the cases when $S_0$ does not have a tiger, we will essentially classify the possible fences $(S_2, A_2 + B_2)$, see (17.3), as well as the original $S_0$, see (17.5-14); we will stop the analysis once we have enough information to apply the criteria of §6, but further analysis of the same sort would yield an explicit classification. The fences $(S_2, A_2 + B_2)$ are in the classification of (13.5), with one exception, (17.3.2), which is partially classified in the proof of (17.3).

**We assume throughout §17 that $S_0$ does not have a tiger, and $\pi_1^{\text{alg}}(S^0) = \{1\}$.**

$K_{T_1} + a_1 E_1$ is flush by (8.4.7). Configurations on $T_1$ are classified in (11.1). We showed in §15 that there is a tiger if the configuration is I. If $A_1 \subset S_1^0$ then $K_{S_1} + A_1$ is trivial and we have a tiger, a contradiction. Thus $A_1$ contains a singular point, and the configuration on $T_1$ is II or higher. Note also that $S_1^0$ is simply connected by (7.3.1).

**17.1.0 Notation:** *Let $C$ and $D$ be the two branches of $A_1$ at the node, and $c$, $d$ the points of $T_1$ where the branches meet $E_1$. We may assume (by switching $C$ and $D$ if necessary) that the first two blow ups of $h : \tilde{T}_1 \longrightarrow S_1$ are along $C$. Let $r+1$ $(r \geq 1)$ be the number of initial blow ups along $C$. Note $d$ is necessarily singular.*

**17.2 Lemma.**

(1) *$K_{T_1} + \Sigma_1 + E_1$ is log canonical.*

(2) *$\overline{\Sigma}$ has two smooth branches through $x_0$ and meets no other singularities.*

(3) *If $c$ is smooth, then $d$ is an $A_r$ point, $r \geq 1$ and $a_1 = \frac{r+1}{r+2}$.*

(4) *$T_1$ is singular at some point of $E_1 \setminus E_1 \cap \Sigma_1$.*

(5) *$S$ has exactly two non Du Val points and $e_0 > 1/2$.*

(6) *$a_1 \geq 2/3$, and $a \geq 4/5$, unless we have (3) with $r \leq 2$.*

(7) *$A$ contains exactly one singularity.*

*Proof.* (1)-(3) are immediate from (11.1.1) (since we have ruled out configurations 0 and I).

(4) holds, for otherwise $A \subset S_1^0$.

For (5), one shows there is at most one non Du Val point away from $A$ as in the proof of (16.3.4). So by (2), there are at most two non Du Val points on $S_0$. Now (5) follows from (10.8).

For (6), suppose we are not in the situation of (3). $a \geq 4/5$ by (3) and (11.0) unless $r = 1$ or $r = 2$. By (11.0) it's enough to consider the configurations after one more blow up. Suppose $r = 1$. After the next blow up $c = (2)$ and $d = (3)$. One computes $a = 4/5$. Suppose $r = 2$. After the next blow up $c = (2)$ and $d = (\underline{2}, 3')$. One computes $a = 6/7$. Hence (6).

Finally suppose (7) fails. Then, since $\pi$ removes one of the singularities along $E$, $A$ contains two singularities, $x_0$ is a non-chain singularity, and we have (3). We will use (8.3.9) to compute $e_0$. If $E_1$ is not a $-2$-curve, then $e_0 \geq \frac{r+1}{r+2}$, a contradiction. Thus we may assume $k = 2$. As $e_0 > 1/2$, (by (5)), $e_0 \geq 2/3$, by the classification of non-chain singularities, and $r \geq 2$.

One checks that $e_0 \geq \frac{r+1}{r+2}$ unless $r = 2$ and the other two singular points are (2) and (3). Thus $\tilde{A}_1^2 = 4$. As $S_1$ has a non Du Val point outside $A_1$, $M$ is reducible by (12.5). Thus $W$ of (12.4) has rank one, and $K_W^2 = \tilde{A}^2 = 4$. $\pi$ (of (12.4)) has two exceptional divisors. $G_1$ and $G_2$ of (12.4) are the $-2$ and $-3$-curves adjacent to $A$, and $G_i \subset Y^0$. By (3.9) $W = S(2A_1 + A_3)$, and each $D_i = \pi(G_i)$ contains two singular points, thus each $G_i$ must meet both $\pi$ exceptional divisors. But then $D_1 \cap D_2$ contains two points, contradicting (3.9). □

By (17.2), there are two non Du Val points, $x_0$ and $y$, of $S_0$ and $S_1$ has exactly one singularity, $z$, along $A$. Let $s$ be the index of $z$. We will first show that $S_2$ is a fence (17.3), and then partially classify each possible $S_0$, and show in each case, using §6, that $S_0^0$ is uniruled. (With a little more analysis one can completely classify the possibilities, but we stop once we have enough information to apply the criteria of §6).

Recall from (8.2.10) that $E_2$ is a $-j$-curve, and $E_1$ is a $-k$-curve.

**17.3 Proposition (Outcome of the hunt).** *$(S_2, A_2 + B_2)$ is a Fence. $g(A_2) = 1$, $B_2$ is smooth.*

(1) *If $x_1 \in A_1$ then $\Sigma_2$ meets $E_2$ at a smooth point, and contains a unique singular point $(A_t, 3, A_{j-2})$ (for some $t$). $K_T + \Sigma_2$ is log terminal, and $\Sigma$ meets the end of the $A_{j-2}$ chain. $(S_2, A_2 + B_2)$ is given by (13.5). $q_2$ is an $A_{t+1}$ point.*

(2) *If $x_1 \notin A_1$, and $A_2 \not\subset S_2^0$, then $S_0^0$ is uniruled.*

(3) *If $x_1 \notin A_1$ and $A_2 \subset S_2^0$, then $(S_2, A_2 + B_2)$ is given by (13.5.1), and $a < 6/7$.*

*Proof.* As $a + e_1 \geq 2/3 + e(y) \geq 1$, we have the same possibilities for the outcome of the hunt as in (16.2).

By (16.6) $T_2$ is not a net.

Suppose $x_1 \in A_1$. Then $S_2 = W$ of (12.4), and $M = \Sigma$. As $M$ is disjoint from $A$ (on $T = Y$ of (12.4)) we must go to a fence, with $B$ smooth by (11.1), since $a_2 + b_2 > 1$. By (13.3) (with $B$ playing the rôle of $Y$ of (13.3)) there are at least two singularities along $B$, and $B$ is a $-1$-curve. Thus $\Sigma_2$ meets $E$ at a smooth point, and contracts to a cyclic Du Val singularity. Now (1) follows easily by (11.4).

Thus we may suppose $x_1 \notin A_1$. Suppose $A_2 \cup B_2$ has a node of genus $g \geq 2$. Since $a_2 + b_2 > 1$, the configuration on $T_2$ is given by (11.1.1.2).

If $z \notin \Sigma$, then $K_{S_2} + A_2$ is not trivial and $b < 1/2$ by (11.1.1.2). Thus $x_1 = (3, A_g)$, by (10.1). But then $B$ is contained in the smooth locus and as $B$ is not a tiger, $g \geq 3$. (11.1.1.2) implies $b \leq 4/9$. On the other hand, $e_1 \geq e(3, A_3) = 4/9$, a contradiction.

If $z \in \Sigma_2$, then $A_2$ is contained in the smooth locus, and $S_2$ is simply connected Gorenstein. As $z = A_g$, $e_1 \geq ag/(g+1)$ and by (11.1.1.2) $2ag < (g+1)$, that is $2/3 \leq a < 1/2 + 1/(2g)$. Thus $a = 2/3$, $g = 2$. By (17.2.6), $r = 1$, and $x_0$ is a chain singularity. In particular $k \geq 3$ by (10.11). Thus

$$S_2^2 = A_2^2 = r + 4 + g - k \leq 4.$$

But $B$ contains a chain singularity (it is not a tiger), which contradicts the list (3.2).

Thus we must go to a fence. $B$ is smooth, and by (13.3) contains at least two singularities, and is a $-1$-curve.

If $A_2 \subset S_2^0$, then $S_2$ is Gorenstein, and $S_2^0$ is simply connected, by (7.3.1), since $q_2$ is smooth. Thus $(S_2, A_2 + B_2)$ is given by (13.5.1). Blowing up the $A_1$ point and contracting $B$ gives a flex line to a nodal cubic, and the inequality $b/2 + 3a < 3$. Since $a > b$, (3) follows.

Now suppose $A_2 \not\subset S_2^0$. Thus $\Sigma_2$ meets $A$ at a smooth point. As $a + e_1 \geq 1$, by (11.1) $\Sigma_2$ meets $E_2$ at an $A_s$ point, for some $s \geq 1$. $E_2$ is a $-2$-curve, so $x_1$ is a non-chain point. Since $A_1$ contains only one singular point, $\tilde{A}_1^2 \geq 1$, by (12.4.6), and so (from the configuration on $T_2$), $\tilde{A}_2^2 = \tilde{A}_1^2 + 1 + s$.

Suppose $\tilde{A}_2^2 \geq 4$ (for example if $s \geq 2$, or $\tilde{A}_1^2 \geq 2$).

We apply (12.4) to $(S_2, A_2)$. Since $A_2$ contains a unique singularity, $W$ has rank one, by (12.4.1). $K_W^2 = \tilde{A}_2^2 \geq 4$, so $W$ is given by (3.9.2). $G$ (of (12.4)) is irreducible, the curve over $z$ adjacent to $A_2$. $M$ is irreducible. $D = \pi(G)$ and $B \subset W$ are each smooth $-1$-curves, and thus their configuration is given by (3.9.2). In particular they meet exactly once, at opposite ends of the $A_3$ point, which is necessarily $\pi(M)$, and each contains in addition an $A_1$ point. Since $G$ contains at most one singularity, $M$ meets $G$ at a smooth point. Now consider the sequence of blow ups $h : \tilde{Y} \longrightarrow \tilde{W}$. The first blow up is along $D$ (since $D$ is a $-1$-curve, and $G$ is $K_Y$ non-negative). All further blow ups must also be along $G$, for otherwise $M$ meets $G$ at a singular point. Say $G$ is a $-v$-curve. Then there is a unique singularity $(\underline{2}, 2, 3, A_{v-2})$ on $M$, at $M \cap B$, $B$ meets the underlined curve, and $K + B + M$ is log canonical (so $M$ meets the end of the $A_{v-2}$

chain). Thus the branches of $x_1$ are $A_s$, (2) and $(2,2,3,A_{v-2})$. Since this last branch has index at least 7, $s \leq 1$. Note also that $z = (\underline{v}, 2)$.

Hence $s = 1$. Then $a_2 + b_2/2 = 1$. By (10.1), $b_2 > e_1 \geq 1/2$, thus $a \leq 3/4$. So by (17.2.6), $a = 2/3$, $r = 1$. Then $b < 2/3$ so $x_1$ is given by (10.1.2.d). As $e(2,4,2) = 2/3$, $k = 3$. Thus $\tilde{A}_1^2 = 4 + r - k = 2$. Thus the singularities are as in the previous paragraph. Since $e_1 = 1/2$, $k = 2$, by (8.0.7.2). Thus $x_0 = (2, 3^A, 2^G, 2)$. The index of $x_0$ is eleven. $\overline{\Sigma}$ has two smooth branches, $(2', 3, 2, 2)$ and $(2, 3', 2, 2)$. One computes $-K_{S_0} \cdot \overline{\Sigma} = 2/11$. Hence $S_0^0$ is uniruled by (6.5). $\square$

**17.4 Lemma.** *Either $x_0$ is a chain singularity, or $S_0^0$ is uniruled.*

*Proof.* Suppose $x_0$ is a non-chain singularity, and $S_0^0$ is not uniruled. We will use (8.3.9) to compute $e_0$.

By (17.2) $c$ and $d$ are singular points.

Suppose $c$ has type (2). By (11.1), the two points of $E_1 \cap \Sigma_1$ are $c = (\underline{2}')$ and $d = (\underline{2}, \dots, 3')$ (where there are $r-1$, $-2$-curves). In this case one checks that $a$ rescales to $\frac{2r+2}{2r+3}$. On the other hand, if $E_1$ is a $-3$-curve, then $e_0 \geq \frac{2r+2}{2r+3}$. Thus $E_1$ is a $-2$-curve. As $e_0 > 1/2$, it follows that the third branch is not an $A_1$-singularity. Thus (by the classification of non-chain singularities) either $r = 1$ and the last point has index at most 5, or $r = 2$, and the last point has index three. One checks the only case is $r = 1$, $e_0 = 2/3$, $\tilde{A}^2 = 3$, $a = 4/5$ and $A$ contains an $A_2$-singularity.

Suppose first $x_1 \notin A_1$. Let $w = \Sigma_2 \cap E_2$. We have (17.3.3). By (11.1), if $w$ is smooth, then since $B$ is a $-1$-curve, $j = 2$ and $x_0$ is Du Val, a contradiction. Thus $w$ is singular, and $x_1$ is a non-chain point with branches (2), (2, 2) and $w$. $z = \Sigma_2 \cap A$. Since $\tilde{A}_2^2 = 6$, the first three blow ups of $h : \tilde{T}_2 \longrightarrow \tilde{S}_2$ are along $A$. Thus by (11.1), $w = (4', \underline{2})$, $w = (4', \underline{2})$, which has index greater than five, a contradiction.

We conclude $x_1 = z$, and $(S_2, A_2 + B_2)$ is given by (17.3.1). $B_2$ contains an $A_1$ point, and $A_2^2 = 3$. It follows from (13.5) that $S_2$ is given by (13.5.2). Let $R'$ be given by (17.5.2). Then $R'$ has a node at $x_0$. One computes each branch has index nine and $-K_{S_0} \cdot R' = 1/3 > 2/9$. Hence $S_0^0$ is uniruled by (6.5).

Now suppose $c$ does not have type (2). Then both $c$ and $d$ have index at least 3, and the indices of singularities along $E_1$ are $2, 3, n$ with $n \leq 5$. Consider the possibilities for $\tilde{T}_1 \longrightarrow \tilde{S}_1$, classified in (11.1). Index consideration shows $r = 1$, and we are exactly one blow up beyond $c = (2)$ and $d = (3)$. There are two possibilities, $c = (\underline{2}, 2')$, $d = (4)$, $a = 6/7$ or $c = (3)$, $d = (\underline{3}, 2')$, $a = 7/8$. But if $x_0$ has centre a $-2$-curve and branches (2), (2, 2) and (4), then $e_0 = 6/7$ and so the former case is impossible. $e_0 \geq 16/17$, in the latter case, again impossible. $\square$

Thus from now on we will assume $x_0$ is a chain singularity. In particular $k \geq 3$, and there are exactly two singularities along $E_1$, $z$ and $A_r$.

### 17.5 Proving $S_0^0$ is uniruled

We now classify possibilities for $S_0$ leading to $(S_2, A_2 + B_2)$ given in (17.3.1) and (17.3.3). In each case we show that $S_0$ is uniruled using the methods of §6. This involves finding certain rational curves on $S_0$. One candidate, by (17.2.2) is $\overline{\Sigma}$. The others we will use are given by the following lemmas. We encourage the reader to first go through example (6.10), as the analysis to follow is very similar, but less detailed.

The notation for the rational curves below is fixed for the remainder of §17.

**17.5.1 Lemma.** *Let $A \subset \mathbb{P}^2$ be a nodal cubic, with node at $q$ and a flex point at $p$. Let $r \neq p$ be the unique point of $A^0$, whose tangent line meets $p$.*

(1) *There is a nodal cubic $R$ through $p$ and $q$, smooth at $p$ and $q$, with seventh order contact with $A$ at $p$.*
(2) *There is a nodal cubic $\overline{R}$ through $p$, $q$, and $r$, smooth at these points, with forth order contact with $A$ at $p$, and third order contact with $A$ at $r$.*

*Proof.* We prove the first case, and the leave the second to the reader, which is analogous. Let $S = S(A_1 + E_7)$. Blow up once at $q$ and seven times at $p$ along $A$, to obtain $\tilde{S}$. It is enough to show there is a nodal elliptic curve $R \subset S^0$. For this we study the elliptic pencil $| - K_S|$. [17] implies there is a smooth member $C \in | - K_S|$, with $C \subset S^0$, and if $V$ is the $-1$-curve over $q$ then $V \in | - K_S|$.

Thus we have an elliptic fibration $\pi : T \longrightarrow \mathbb{P}^1$ after blowing up the basepoint of $| - K_S|$. Then by (16.4) of [4] we have

$$e(\tilde{T}) = e(\mathbb{P}^2) + 9 = e(I_2) + e(\tilde{E}_7) + n$$

where $e$ is the Euler characteristic, $n$ is the contribution to the sum from the other singular fibres and $\tilde{E}_7$, $I_2$ are as on page 150 of [4] (The fibre through the $A_1$ point contains $V$ and so we get the graph $I_2$). Thus $n = 1$ and one fibre of $\pi$ is a nodal cubic $R$. □

*17.5.1.1 Remark.* By a dimension count one expects in (17.5.1.1) a one dimensional family of elliptic curves passing through $p$ and $q$ meeting $A$ to order 7 at $p$. However finding a *rational* elliptic curve in the family, smooth at $p$ and $q$, is more subtle. For example the proof of (17.5.1) shows that there is no such curve if $R$ has an ordinary cusp.

**17.5.2 Lemma.** *Let $A$, $C \subset \mathbb{P}^2$ be a nodal cubic, with node $q$, and a smooth conic, meeting $A$ to order six at a point $p \neq q$. Then there is a nodal cubic $R'$ which meets $A$ at $p$ to order seven.*

*Proof.* If we blow up 3 times at $p$ along $A$ and then blow down the tangent line and the two $-2$-curves, then $C$ becomes the flex line to $p$, and the result follows from (17.5.1.2). □

**17.5.3 Lemma.** *Let $A$, $C \subset \mathbb{P}^2$ be a nodal cubic, with node $q$, and a smooth conic, meeting to order six at a point $p \neq q$. Then there is a rational quartic $R''$, with an oscnode at $p$ (that is two smooth branches meeting to order 3), such that one branch of $R''$ at $p$ meets $A$ to order 5, and the other meets $A$ to order 3, and such that $R''$ meets one branch of $A$ at $q$ to order 3.*

*Proof.* After a birational transformation (cf. the proof of (17.5.2)) $p$ becomes a flex point and the problem is to find a smooth conic $Q$ tangent to $A$ at $p$ with the described contact at $q$. As all pairs $(A, C)$ are projectively isomorphic, it is enough to construct $Q$ for some pair.

Begin with a smooth conic $E \subset \mathbb{P}^2$, two tangents $C$ and $G$ at $c$ and $g$, and a secant line $A$ through $c$, $g \notin A$. Let $w : T \longrightarrow \mathbb{P}^2$ be obtained by blowing up 3 times at $c$ along $C$, and 3 times at $E$ along $g$. Let $Q$ be the $-1$-curve over $g$, and $V$ the $-1$-curve over $c$. Let $h : T \longrightarrow \mathbb{P}^2$ be given by contracting all the $w$-exceptional curves except $Q$ and $V$, together with $E$ and $G$ (which are both $-1$-curves on $T$). $Q$ is the required conic. □

**17.5.4 Definition.** *Suppose $(S_2, A_2 + B_2)$ is given by (13.5.1). Let $Z, W \subset S$ be the strict transforms of the tangent lines to the two branches $C$, $D$, of (17.1.0), of $A \subset \mathbb{P}^2$ at the node $q$. Note $W$ has a unibranch singularity at $x_0$ of form $(A_{r-1}, 2', k', \dots)$ (on $\tilde{T}_1$, $\tilde{W}$ meets the intersection of $\tilde{E}_1$ with the first $-2$-curve of the $A_r$ point). Note if $r \geq 2$, $Z$ is smooth at $x_0$, and has form $(A_{r-2}, 2', 2, k, \dots)$, while if $r = 1$, $Z$ meets $\Sigma_1$ at a smooth point, and does not pass through $x_0$.*

Note $-k + r + 4 = \tilde{A}_1^2$. We have

$$0 > (K + aA_1) \cdot A_1 = \frac{s-1}{s} - \frac{A_1^2}{r+2} \tag{17.6 }$$

Suppose $z$ has type $\Gamma_n$. Then $A_1^2 = \tilde{A}_1^2 + \frac{s-1-n}{s}$ and one finds from (17.6)

$$\tilde{A}_1^2 > (s-1)(k-3) + n \tag{17.6.1}$$

In 17.7-12 we classify possibilities where $x_1 \in A_1$. (17.3.1) describes the configurations on $S_2$ and $T_2$. We consider each possibility for $S_2$ in turn. In any case $s \geq 3$. Note $\tilde{A}_1^2 = K_{S_2}^2$.

**17.7** $S_2 = S(A_1 + A_2)$. Then $\tilde{A}_1^2 = 6$, $r = k + 2$. Note $E_2$ has either an $A_1$ point or $A_2$ point along it.

**17.7.1 If $E_2$ has an $A_2$ point.** Thus $z = (j, 2, 2)$, $y = (A_{j-2}, 3)$, $x_0 = (2, 2, j, k, A_{k+2})$. $s = 3j - 2$. $n = 3j - 6$. By (17.6.1), $j < 4$.

**17.7.1.1 If** $j = 3$. (17.6.1) gives $k = 3$. $\Delta(x_0) = 73$. $-K_{S_0} \cdot \overline{\Sigma} = 3/73$. Log uniruled by (6.5). □

**17.7.1.2 If $j = 2$.** (17.6.1) gives $k = 3$, 4, $\Delta(x_0) = 34, 67$. One checks the smooth branch of $\overline{\Sigma}$ has index 17 in the first case. $-K_{S_0} \cdot \overline{\Sigma} = 6/34, 3/67$. Log uniruled by (6.5). □

**17.7.2 If $E_2$ has an $A_1$ point.** Thus $z = (j, 2)$, $y = (A_{j-2}, 3, 2)$, $x_0 = (2, j, k, A_{k+2})$. $s = 2j-1$, $n = 2j - 4$. (17.6.1) gives $j < 5$.

**17.7.2.1 If $j = 4$.** (17.6.1) gives $k = 3$. $\Delta(x_0) = 79$. $-K_{S_0} \cdot \overline{\Sigma} = 2/79$. □

**17.7.2.2 If $j = 3$.** (17.6.1) gives $k = 3$. $\Delta(x_0) = 53$. $-K_{S_0} \cdot \overline{\Sigma} = 4/53$. □

**17.7.2.3 If $j = 2$.** (17.6.1) gives $k = 3, 4, 5$. $\Delta(x_0) = 27$, 52, 83. Note the two branches of $Z$ at $x_0$ are $(2', j, k, A_{k+2})$ and $(2, j, k, 2, 2', A_k)$. One checks the smooth branch of $Z$ has index 27, 52 in the first two cases. $-K_{S_0} \cdot Z = 9/27$, 8/52, 5/83. □

**17.8 If $S_2 = S(A_1 + A_5)$.** Then $\tilde{A}_1^2 = 3$, $r = k - 1$. Note $E_2$ has either an $A_1$ point or $A_5$ point along it.

**17.8.1 If $E_2$ has an $A_5$ point.** Thus $z = (j, A_5)$, $y = (A_{j-2}, 3)$, $x_0 = (A_5, j, k, A_{k-1})$. $s = 6j - 5$, $n = 6j - 12$. (17.6.1) gives $j = 2$, $k = 3$. $\Delta(x_0) = 31$, $-K_{S_0} \cdot \overline{\Sigma} = 3/31$, so log uniruled by (6.5). □

**17.8.2 If $E_2$ has an $A_1$ point.** $z = (j, 2)$, $y = (A_{j-2}, 3, A_4)$, $x_0 = (2, j, k, A_{k-1})$. $s = 2j - 1$, $n = 2j - 4$. (17.6.1) gives $j = 3, 2$.

**17.8.2.1 If $j = 3$.** (17.6.1) gives $k = 3$. $\Delta(x_0) = 29$, $\overline{\Sigma}$ has two smooth branches at $x_0$. $R''$, (of 17.5.3) has two smooth branches at $x_0$. Thus log uniruled by (6.5), using $R''$ and $\overline{\Sigma}$.

**17.8.2.2 If $j = 2$.** (17.6.1) gives $k = 3, 4$. For $k = 3$, $\Delta(y) = 11$. $R''$ has two smooth branches at $y$, $(A_4, 3')$ and $(A_2, 2', 2.3)$. $-K_{S_0} \cdot R'' = 3/11$. log uniruled by (6.5). For $k = 4$, $\Delta(x_0) = 31$. $R'$ has one smooth, and one singular branch at $x_0$. The singular branch is $(2, 2, 4', 2', A_2)$, which one checks has index 31. Thus (with $R'$ and $\overline{\Sigma}$) log uniruled by (6.5). □

**17.9 If $S_2 = S(3A_2)$.** $\tilde{A}_1^2 = 3$. $r = k - 1$. $y = (2, 3, A_{j-2})$, $x_0 = (A_2, j, k, A_{k-1})$. $s = 3j - 2$, $A_1^2 = 3 + \frac{3}{3j-2}$. Then (17.6) gives $j = 2$, $k = 3$. $\Delta(x_0) = 19$, and $-K_{S_0} \cdot \overline{\Sigma} = 3/19$, log uniruled by (6.5). □

**17.10 If $S_2 = S(A_2 + A_5)$.** $\tilde{A}_1^2 = 2$. Thus by (17.6.1), $k = 3$, $r = 1$, $j = 2$, and $z$ is Du Val . Note $E_2$ contains either an $A_2$ or an $A_5$ point.

**17.10.1 If $E_2$ contains an $A_2$ point.** Thus $y = (A_4, 3)$ and $x_0 = (2, 3, A_3)$. $\Delta(x_0) = 14$. One checks the smooth branch of $\overline{\Sigma}$ has index 14, and $-K_{S_0} \cdot \overline{\Sigma} = 2/14$, thus log uniruled by (6.5). □

**17.10.2 If $E_2$ contains an $A_5$ point.** $y = (2, 3)$, $x_0 = (2, 3, A_6)$. $\Delta(x_0) = 23$. $-K_{S_0} \cdot \overline{\Sigma} = 2/23$, thus log uniruled by (6.5). □

**17.11 If $S_2 = S(A_1 + 2A_3)$.** By (13.5.5) $E_2$ contains an $A_3$ point. As in (17.10) $\tilde{A}_1^2 = 2$ implies $k = 3$, $r = 1$, $j = 2$. We have $y = (A_2, 3)$ and $x_0 = (2, 3, A_4)$. $\Delta(x_0) = 17$, $-K_{S_0} \cdot \overline{\Sigma} = 2/17$, so log uniruled by (6.5). □

**17.12 If $S_2 = S(2A_1 + A_3)$.** Note $E_2$ contains either an $A_3$ or an $A_1$ point. We have $\tilde{A}_1^2 = 4$. $r = k \leq 4$ by (17.6.1). Suppose $k = 4$. (17.6.1) gives $n = 0$, and $5 > s$, thus $j = 2$, and $E_2$ contains an $A_1$ point, $y = (A_2, 3)$, $x_0 = (A_4, 4, A_2)$, $\Delta(x_0) = 38$, one checks the smooth branch of $\overline{\Sigma}$ had index 38 and $-K_{S_0} \cdot \overline{\Sigma} = 2/38$, so log uniruled by (6.5).

Now assume $k = r = 3$.

**17.12.1 If $E_2$ contains an $A_3$ point.** By (17.6.1), $n < 4$ implies $j = 2$. Thus $x_0 = (A_3, 3, A_4)$, $\Delta(x_0) = 29$, $-K_{S_0} \cdot \overline{\Sigma} = 4/29$, log uniruled by (6.5). □

**17.12.2 If $E_2$ contains an $A_1$ point.** $z = (\underline{j}, 2)$, which has type $2j - 4$. Thus (17.6.1) gives $j = 2, 3$. For $j = 2$, $x_0 = (A_3, 3, A_2)$, $\Delta(x_0) = 19$, $-K_{S_0} \cdot \overline{\Sigma} = 4/19$. For $j = 3$, $x_0 = (A_3, 3^A, 3, 2)$, $\Delta(x_0) = 37$, $-K_{S_0} \cdot \overline{\Sigma} = 2/37$. □

In 17.13-14 we classify possibilities where $x_1 \notin A_1$. Thus we have the setup of (17.3.3). We divide the analysis into two cases depending on whether or not $\Sigma_2 \cap E_2$ is a smooth point. Note $y = x_1$.

**17.13 If $\Sigma_2$ meets $E_2$ at a smooth point.** So $z$ is a Du Val point, $A_t$, $j = t + 2$, $\tilde{A}_1^2 = 5$, $r = k + 1$,$x_0 = (A_t, k, A_{k+1})$, $x_1 = (2, t + 2, 2, 2)$. $s = t + 1$, so (17.6.1) gives

$$5 > t(k - 3) \tag{17.13.1}$$

In particular $k < 8$.

**17.13.2 If $k = 3$.** One checks $t = 1$ (or $e_1 > e_0$). $x_0 = (2, 3, A_4)$, $\Delta(x_0) = 17$. $-K_{S_0} \cdot \overline{\Sigma} = 5/17$. $S_0$ is log uniruled by (6.5). □

**17.13.3 If $k = 4$.** 16.16.1 give $t = 1, 2, 3, 4$. $\Delta(x_0) = 32, 45, 58, 71$. One checks in each case that the singular branch of $R$, which is $(A_{t-1}, 2', 4^3, A_5)$ has index $\Delta(x_0)$, and that the smooth branch, which is $(A_t, 4, 2', A_4)$, has index $\Delta(x_0)$ except in the case $t = 2$ when the index is 9. One computes $-K_{S_0} \cdot R = \frac{6(5-t)}{13t+19}$. Thus $S_0$ is log uniruled by (6.5) in every case. □

**17.13.4 If $k = 5$.** 16.16.1 gives $t = 1, 2$. $\Delta = 51, 73$. One checks in the first case that the smooth branch $\overline{\Sigma}$ has index 51, and $-K_{S_0} \cdot \overline{\Sigma} = 3/51$. In the second case $R$ has two branches at $x_0$, one smooth and one singular. One checks the singular branch has index 73. The singular branch is $(A_6, 5^3, 2', 2)$. The smooth branch is $(A_5, 2', 5, 2, 2)$. One computes $-K_{S_0} \cdot R = 7/73$. So log uniruled by (6.5). □

**17.13.6 If $k = 6$. $t = 1$.** $\Delta(x_0) = 74$. One checks both branches of $R$ have index 74. $-K_{S_0} \cdot R = 16/74$. □

**17.13.7 If $k = 7$. $t = 1$.** $\Delta(x_0) = 101$. One checks the singular branch of $R$ has index 101. $-K_{S_0} \cdot R = 9/101$. □

**17.14 If $\Sigma_2$ meets $E_2$ at a singular point.** Let $w = E_2 \cap \Sigma_2$. $x_1$ is necessarily a non-chain point, with branches $w, (2), (2,2)$. From an easy classification (since $x_1$ is not Du Val), $e_1 \geq 2/3$. Consider $h : \tilde{T}_2 \longrightarrow \tilde{S}_2$.

**17.14.1 If the second blow up of $h$ is along $A$.** Note $j = 2$. The only possibility for $w$ (with $e_1 < 6/7$, and $x_1$ non Du Val) is (3). Then $z = (2), \tilde{A}_1^2 = 4$. $R$ (on $\tilde{S}_0$) is a 3-curve. Its singular branch at $x_0$ is $(2', k^2, A_k)$ and its smooth branch is $(2, k, 2', A_{k-1})$. Also after scaling $b = 2 - 3/2a$. Then from the configuration in $\mathbb{P}^2$ it follows that $a < 8/9$. Thus $r = k \leq 6$. If $k = 3$ then $e_0 < e_1 = 2/3$. Thus $k = 4, 5, 6$. $\Delta(x_0) = 27, 44, 65$. For $k = 4$ one checks the smooth branch of $\overline{\Sigma}$ has index 27, and $-K_{S_0} \cdot \overline{\Sigma} = 3/27$. For $k = 5$ one checks both branches of $R$ have index 44 and $-K_{S_0} \cdot R = 12/44$. For $k = 6$ the singular branch of $R$ has index 13, the smooth branch has index 65, while $-K_{S_0} \cdot R = 7/65 > 1/13 + 1/65$. So (6.5) applies in each case. □

**17.14.2 If the second blow up of $h$ is along $E_2$.** Note $j \geq 3$. If $j \geq 4$ then $e_1 \geq 6/7$. Also if $j \geq 3$ and $w$ has index at least 3, $e_1 \geq 6/7$. We conclude $w = (2)$, $j = 3$, $z = (3)$, $\tilde{A}_1^2 = 5$, $r = k + 1$, $e_1 = 3/4$, $x_0 = (3, k, A_{k+1})$. By (17.6.1) $k \leq 4$. If $k = 3$ then $e_0 = 21/28 < 3/4$, a contradiction. So $k = 4$. $\Delta(x_0) = 51$. $R$ has branches $(3', 4^3, A_5)$ and $(3, 4, 2', A_4)$. One checks each has index 51. $-K_{S_0} \cdot R = 4/17$. □

## §18 $A_1$ SMOOTH

In this and the following section we prove

**18.1 Proposition.** *If $A_1$ is smooth, then either $S$ has a tiger, or $(S_2, A_2 + B_2)$ is a smooth banana, classified in (13.2). In either case $S_0^0$ is uniruled.*

**We assume throughout §18–19 that $S$ has no tiger.**

In this section we will consider all possibilities for the next hunt step, and rule out all but the smooth banana. The smooth banana will be considered in §19

Recall from (8.2.10) that $E_1$ is a $-k$-curve, and $E_2$ is a $-j$-curve. We first put together some elementary observations.

$K + aA$ is flush and $K_{S_1} + A$ is lt by (8.4.7).

**18.2 Lemma.** *The following hold:*

(1) *$A$ is ample and $\tilde{A}^2 \geq -1$.*

(2) *A contains exactly three singularities.*

(3) *There is a singularity of index at least four along $A$.*

(4) $a > 2/3$, $e_1 > 1/2$.

*Proof.* (1) follows, since $K_{S_1} \cdot A < 0$.

Next we consider (2). As $A$ is not a tiger, $A$ must contain at least three singularities, by adjunction. Suppose it contains at least 4. Since $E$ can contain at most three, and the only additional singularity can be at $q_1$, we conclude $x_0$ is a non-chain singularity, $\Sigma \cap E$ is a smooth point, and there are exactly four singularities along $A$. By adjunction one has index at least three.

$e_0 \geq 1/2$ by (10.1). $e_1 \geq 2/3a$ (since there is a singularity along $A$ of index at least three), by (8.3.8). If $e_0 \geq 2/3$, then $a_2 + 3b_2 > 2$. If $e_0 < 2/3$, then $x_0$ is given by (10.1.d), and $A$ contains a non-Du Val singularity, so $e_0 \geq e_1 \geq 1/2$ by (8.0.7). Thus again $a_2 + 3b_2 > 2$. In particular $a_2 + b_2 > 1$.

Consider the next step of the hunt; the possibilities are given by (8.4.7).

Suppose $T_2$ is a net. $A$ is not fibral by (11.5.5). $A+E$ has degree at least three, for otherwise $-(K_T + A + E)$ is nef and we have a tiger. Since $a_2 + 3b_2 > 2$, it follows that one of $A$ or $E$ is a section, and the other is a double section. This contradicts (11.5.11.2). Thus $\pi_2$ is birational.

The possibilities for $(S_2, A_2 + B_2)$ are given by (8.4.7). $A_2$ must contain at least two singularities. Thus $B$ is not smooth, by (13.1). Hence we have a fence. $A_2$ contains three singularities, thus by (13.3), $S_2 = S(A_1 + A_2 + A_5)$, $B \subset S_2^0$, $B$ has arithmetic genus one, and a simple node, and $B^2 = 1$. By (11.1.1), $T_2$ is smooth along $\Sigma_2 \setminus \Sigma_2 \cap E_2$, and $\Sigma \cap B$ is an $A_r$ point, and

$$1 = B_2^2 = r + 1 - j.$$

Thus $r = j$. $x_1 = (\underline{j}, A_j)$. $S_1$ is smooth away from $A_1$, and its singularities are exactly $x_1, A_1, A_2, A_5$. $S_1$ cannot be Gorenstein (since $\tilde{S_1}$ has Picard number at least nine). One of the singular points is $q_1$. This cannot be $x_1$(or $x_0$ has branches of index $(2,3,6)$). Thus $x_1$ has index at most five, either $x_1 = (3)$, $x_1 = (4)$ or $x_1 = (3,2)$. This contradicts our expression $x_1 = (j, A_j)$. Hence (2).

(3) follows immediately from (2) and adjunction.

Now we prove (4). The second inequality follows from the first, and (3) by (8.3.8).

Suppose first $x_0$ is a chain singularity. Then $\Sigma$ meets $A$ at a smooth point and $q_1$ is a singular point. The configuration on $T$ is of form $(n^s)$ of (11.4). There is a unique singularity, $y$, along $\Sigma$, and $K_T + \Sigma$ is lt. $k \geq 3$, so the self-intersection of $E$ goes up by at least two, and thus $y$ is of the form $(2', s, \dots)$ (for some $s \geq 3$) where $x_1$ has type $(\underline{s-1}, \dots)$. Let $g$ be the coefficient of the $-s$-curve. Then the marked $-2$-curve has coefficient $g/2$, and $a + g/2 = 1$. $K_T + aE$ is flush by (8.4.7), thus $a > g$, so $a > 2/3$.

Now assume $x_0$ is not a chain singularity, and $e_0 < 2/3$. Then by (10.1.1.d), two of the singular points along $E$ are index two, $E$ is a $-2$-curve, and $e = 1/2$. We may assume $\Sigma \cap E$ is one of the (2) points, for otherwise the indices along $A$ are $(2, 2, m)$ for some $m$, and we have a tiger by adjunction. The configuration on $T$ is $(n^s, f)$ of (11.4), for some $s \geq 1$ (in no other configuration does $\Sigma$ meet $E$ at an $A_1$ point). There is one singularity $y$ along $\Sigma \setminus A \cap \Sigma$. Let $g$ be the coefficient of the curve over $y$ which meets $\Sigma$. We have $a/2 + g = 1$. Arguing as before, $a > 2/3$. Hence (4). □

Define $t > 0$ so that $K_{S_1} + tA$ is numerically trivial. $K_{S_1} + tA$ is flush by (8.3.2.1) and (8.3.5.2).

**18.3 Lemma.** *Suppose $A$ is a $(l-2)$-curve, and $A$ has only Du Val singularities along $A$. If $\epsilon = (K_S + A) \cdot A$, then*

(1) $t = l/(l+\epsilon)$, $\epsilon = l(1-t)/t = A^2 - l$

(2) *$S_1$ is not Gorenstein.*

*Proof.* We have

$$\begin{aligned} 0 = (K_S + tA) \cdot A &= K_S \cdot A + tA^2 \\ &= -l + t(l+\epsilon) = -l + lt + t\epsilon \end{aligned}$$

(1) follows.

Suppose $S_1$ is Gorenstein. By (18.2.2), $A$ contains three Du Val chain singularities, say of index $s+1$, $v+1$, $r+1$. By adjunction

$$\frac{s}{s+1} + \frac{v}{v+1} + \frac{r}{r+1} > 2.$$

One checks from the list (3.1) that no rank one Gorenstein log del Pezzo contains three such singularities. □

*18.4 Remark.* Since $a_1 + e_1 > 1$, by (8.4.7) one of the following occurs.

(a) $(S_2, A_2 + B_2)$ is a tacnode.

(b) $S_2$ is a fence.

(c) $S_2$ is a banana.

(d) $A$ is contracted, that is $A = \Sigma_2$.

(e) $T_2$ is a net.

Furthermore $K_{T_2} + \Gamma'$ is flush, and $K_{S_2} + a_2 A_2 + b_2 B_2$ is flush in cases (a-c).

**18.5 Lemma.** *(18.4.a) never holds.*

*Proof.* Suppose (18.4.a) holds. By (11.1.1) the configuration on $T_2$ is type $II$, and $2a_2+3b_2 \leq 3$.

Suppose $S_1$ has a non Du Val Point along $A$. Then by (8.3.8)

$$b_2 > e_1 \geq (1/r) + (r - 2/r)(2/3)$$

and $r > 2$, where $r$ is the index of this point. Thus

$$2a_2 + 3b_2 > 2(2/3) + (2r-1)/r = 2 + 4/3 - 1/r \geq 3,$$

a contradiction.

So $S_1$ is Du Val along $A$. Let $w = e(E_2, K_{S_1} + tA)$. Then $t > w > e_1$, and $K_{T_2} + tA + wE$ is $\pi_2$-trivial, so $2t + 3w \leq 3$ as in (11.1.1.2). Thus $t < 3/4$ by (18.2.4). By (18.3.1), $\epsilon > l/3 \geq 1/3$. Applying the Bogomolov bound, we deduce that either $S_1$ is smooth away from $A$, or it has exactly one singular point away from $A$, a point of index 2 (the contribution to the sum in (9.2) for the singularities away from $A$ is less than 2/3). But then $S_1$ is Gorenstein, contradicting (18.3.2). □

**18.6 Lemma.** *(18.4.b) does not hold.*

*Proof.* Suppose (18.4.b) holds. By (13.3), $B_2$ is singular, and $A_2$ is a $-1$-curve. Since $B_2$ is singular, $x_1 \in A_1$. By (8.0.7.1) and (8.0.4), there is at most one singularity of $S_2$ along $B_2$. By (13.4) and (18.2.4), if $B$ has genus one, then $B$ has a node.

Suppose $A_2$ lies in the Du Val locus. Then $(K+B)\cdot A_2 = 0$, so $B \subset S_2^0$, $B$ has a genus one, and $S_2$ is Gorenstein. By (11.1.1) we have a contraction of type (II, $x^{r-1}$). As $S_1$ is not Gorenstein, by (18.3.2), $x_1$ has type $(j, A_r)$, $j \geq 3$, which has spectral value at least $r+1$. By (8.0.7) this contradicts (11.1.1.3).

Thus we may assume $A_2$ has at least one non Du Val point along it. Thus $e_1 > 1/2$ by (8.0.7). If $B$ has genus one, then it has a node and does not lie in the smooth locus (or $S_2$ is Du Val). Thus $\Sigma_2$ meets $E$ at smooth points, and we have a node of type I. This contradicts (11.1.1.1).

Thus the genus of $B$ is $g \geq 2$. By (11.1.1.5) and (11.1.1.12), and (8.0.7) the spectral value of $\Delta_1$ is at most one. Thus by (8.0.8), $A_1$ has Du Val or almost Du Val points along it. We consider the possible types of contraction, see (11.1) and (11.2).

By (18.2.4), $e_1 > 1/2$, so we cannot have a contraction of type I. Suppose we have one of type II. Then

$$e_1 < b_2 \leq (g+1)/(2g+1) \leq 2/3$$

and so the point we blew up must be an $A_{g+1}$ (otherwise $x_1$ has spectral value at least two). But $A$ has a singularity with spectral value one. Since we still decided to blow up the $A_{g+1}$

point, we have that $a$ is at least the solution to the following equation in $x$

$$1/3 + x/3 = x(g+1)/(g+2).$$

Thus $a \geq (g+2)/(2g+1)$ and $e_1 \geq (g+1)/(2g+1)$, a contradiction.

It follows by (11.1.1) that $B_2$ has a unibranch singularity. Now type III is impossible, as $S_1$ would violate the Bogomolov bound (9.2).

This only leaves types $u$ and $v$.

In type $u$, $x_1 = (\underline{j}, 2, 2)$, and there is in addition a point $(4,2)$ (index 7) away from $A_1$. By spectral value, (8.0.8), $j = 2$.

In type $v$ there is a $(2,3,2,2)$ point (index 11) away from $A_1$.

Now consider the possible indices of the singular points along $A_1$. Let $p$ be the largest index of a non Du Val point along $A_2$ (we have already noted there is at least one non Du Val point along $A_2$), let $q$ be the index of the other point on $A_2$. By (8.3.8) the index of $x_1$ is at least $p$ (or the hunt would not choose $x_1$).

Suppose $q$ is at least three. Then we have four singularities of index at least three on $S_1$ and so we may use the list (10.5). Since we have a point off $A_1$ of index at least 7, it follows from the list that $p$ and $q$ both have index exactly three. But then since $A_2$ is a $-1$-curve, $A_2^2 \leq 0$, a contradiction.

Thus $q$ has index two. As $A_2$ is not contractible, $p$, and hence $x_1$, have index at least five. Thus we have configuration $u$ (in $v$ the index of $x_1$ is four) and $S_1$ has singularities of index at least $2, 5, 5, 11$, which violates (9.2). □

**18.7 Lemma.** *$A$ is not contracted.*

*Proof.* Suppose $A$ is contracted, in other words $A = \Sigma_2$.

By (8.3.1.2) and (8.3.5.1), $q_2$ is a smooth point of $S_2$, $B_2$ has a unibranch singularity and $\Sigma_2$ is the only divisor whose coefficient in $K_{S_2} + bB$ is greater than $b$. Now $\Sigma_2$ cannot be in the Du Val locus, or $B_2$ would be smooth. Thus $e_1 > 1/2$.

Suppose $B_2$ has a triple point. Note that the first exceptional divisor lying over $q_2$ has coefficient $3b - 1 = (2b-1) + b > b$. Thus $\Sigma_2$ must be this divisor, and in particular $\Sigma_2$ lies in the smooth locus, a contradiction.

Thus $B_2$ has multiplicity two at $q_2$. Since $T_2$ has exactly two singularities along $\Sigma_2$, and meets $E_2$ at a smooth point, we must have type III of (11.2).

Suppose $g \geq 2$. Then $b > 2/3$, since $A_1$ has spectral value at least two. But if we blow up $q_2$ twice then the resulting exceptional divisor, which is not $\Sigma_2$, has coefficient $2(2b-1) = b + (3b-2) > b$, a contradiction.

Thus $g = 1$. Exactly as in the proof of (16.3.1), we see that $S$ has a tiger, a contradiction. □

**18.8 Lemma.** *$\pi_2$ is not a net.*

*Proof.* Suppose $\pi_2$ is a net. Let $d$ be the degree of $E_2$ over $\mathbb{P}^1$.

Suppose $A$ is a fibre of $\pi_2$. $d > 2$ (or $K_{T_2} + A + E_2$ is anti-nef) and so $e_1 < 2/3$. Thus $A$ and $E$ have singularities of spectral value at most one. $A$ contains two singular points of $T_2$. Since $e_1 \geq 1/2$, $d = 3$. It follows that $A$ is given by (11.5.9.5) with $k = 2$. Thus by (18.2.3) $x_1$ has index at least 4. Since the spectral value is at most 1, $E$ must be a $-2$-curve, and contain a unique singularity $z$, either Du Val or almost Du Val. Let $h : \tilde{T}_2 \longrightarrow W$ be a $K_T$ relative minimal model, an isomorphism at the generic point of the $-3$-curve meeting $A$. $h$ is determined by selecting which reduced component of each multiple fibre becomes the fibre of $W$, see (11.5.5).

$A$ and $G$, the fibre through $z$, are the only multiple fibres: $E \longrightarrow \mathbb{P}^1$ is necessarily ramified over the image of any multiple fibre, in particular it is ramified at $z$, and totally ramified at $E \cap A$. If there were another multiple fibre, then it would meet $E$ at a smooth point, and so $E$ would be totally ramified there as well, a contradiction.

We consider in turn the two possibilities, either $E$ meets $G$ only at $z$, or at $z$ and also at some smooth point $y$.

In the first case $E$ meets a curve over $z$ of multiplicity 3 in the fibre. From (11.5.9) the possibilities for $G$ are (5), with $k = 3$, or (6). Choose $h$ by selecting in the first case the $-4$-curve, and in the second the $-3$-curve over $z$. Then $h(E) \subset W$ is a smooth rational triple section, and $h(E)^2 = 4$, contradicting (11.5.10).

In the second case, necessarily $m(G) = 2$. The only possibility is (11.5.9.1) (in (5) with $k = 1$, $x_1$ has index 3). Choose $h$ by selecting the $-2$-curve meeting $E$ at $q$. $h(E) \subset W$ is a smooth triple section of self-intersection 4, contradicting (11.5.10).

Thus $A$ dominates $\mathbb{P}^1$. Let $\delta$ be the degree of $A$ over $\mathbb{P}^1$. $d + \delta > 2$ (otherwise $K_{T_2} + A + E$ is anti-nef and we have a tiger). By (18.2.4), one of $E$ or $A$ is a section and the other is a double section. Thus $x_1 \in A$ by (11.5.11.2). The spectral value of $x_1$ is either one or zero.

Suppose $E$ is a section. Each singular point of $A$ lies on a different multiple fibre, by (11.5.5), since each must meet $E$ at a singular point. Thus there are at least two multiple fibres, and so at least two singularities along $E$, a contradiction. Thus $A$ is a section, and $E$ is a double section. There are exactly two multiple fibres by (11.5.11.1).

Suppose $E \subset T^0$. Then $E$ is a $-3$-curve (by spectral value $j \leq 3$, and the hunt would not choose an $A_1$ point as there is a point of index 4 along $A$, by (18.2.3)). Both multiple fibres are of multiplicity two, each given by (11.5.9.1) (if either is (11.5.9.5), with $k = 1$, then the indices of $S_1$ along $A$ are at most (2,3,4) and $K_{S_1} + A$ is negative, a contradiction). Then $h(E) \subset W$, of (11.5.4), is a smooth double section of self-intersection 3, contradicting (11.5.10). Thus there is a unique singular point, $z$, along $E$, and $E$ is a $-2$-curve.

Let $F_1$, $F_2$ be the two multiple fibres of (11.5.11). Each meets $A$ at a singular point. One of the $F_i$, say $F_2$ meets $E$ at $z$. $E$ meets a curve of multiplicity two at $z$. By spectral value (with respect to both $A$ and $E$) the only possibility, from (11.5.9) is

$$(-\underline{3}) + 3(-1) + 2(-\underline{2}) + (-2).$$

Thus $x_1 = A_3$, and one other singularity of $A$ is (3). Thus the last singularity has index at least 3. $F_1$ has multiplicity two. Since it meets $A$ at a point of index at least 3, the only possibility from (11.5.9) is (1), and so the third singularity along $A$ is $(\underline{2}, 2, 2)$. Now if we choose $h : \tilde{T}_2 \longrightarrow W$ an isomorphism along $A$, $h(E)$ is a smooth double section of self-intersection 3, violating (11.5.10). □

## §19 THE SMOOTH BANANA

Here we prove:

**19.1 Proposition.** *If $S_2$ is a smooth banana then $S_0$ is log uniruled.*

**We assume throughout §19 that $S_0$ does not have a tiger.**

Smooth bananas are classified in (13.2), where we find $S_3 = \overline{\mathbb{F}}_2$, $x_2 = (\underline{r+1}, 2)$. We will explicitly classify the possibilities. Note $\Sigma_1$ meets $E_1$ at a smooth point, and so by (11.1.1) meets $A_1$ at a Du Val point, say $A_s$.

$A + B + C \subset \overline{\mathbb{F}}_2$ is a configuration of two sections and a fibre, thus since $K_{\overline{\mathbb{F}}_2} + aA + bB + cC$ is negative

$$(*) \qquad 2a_2 + 2b_2 + c < 4.$$

In particular $e_2 < 4/5$, so by (8.0.7.2), $x_2$ has spectral value at most three. Thus $r \leq 2$.

**19.2 Lemma.** *There are two possibilities:*
**Case 1:** *$x_2 \in B$, the singularities of $S_1$ along $A_1$ are $x_1 = (\underline{s-1}, r+1, 2)$, $A_s$, $A_r$, $s \geq 3$, and $A_1$ is a 0-curve. $(s, r)$ is either $(3, 2)$ or $(4, 1)$.*
**Case 2:** *$x_2 \in A$, the singularities along $A_1$ are $x_1 = (\underline{s}, A_r)$, $A_s$, $(\underline{r+1}, 2)$, $A_1$ is a 1-curve $(s \geq 2)$. $(s, r)$ is either $(3, 2)$, $(4, 1)$ or $(4, 3)$.*

*Proof.* The descriptions of the singularities follow from (11.1.1) and our description of $(S_3, A + B + C)$. Thus we need only determine $s$ and $r$. We compute the spectral value of $x_1$ and use $(*)$:

In Case 1: Suppose first $r = 1$. If $s = 3$ then $K_{S_1} + A_1$ is numerically trivial and we have a tiger. If $s \geq 5$ then $x_1$ has spectral value at least 6, and so $a_2 > b_2 \geq 6/7$ and $c \geq 2/3 \cdot 6/7$, which contradicts $(*)$. If $s = 3$ then $K_{S_1} + A$ is numerically trivial, and $A$ is a tiger.

Now suppose $r$ is two, we show $s$ is at most 3. If $s \geq 4$ then $x_1$ has spectral value at least 7, $a_2 > b_2 \geq 7/8$, and $c \geq 4/5 \cdot 7/8$ which again violates $(*)$.

In Case 2: Suppose first $r = 2$. $s$ cannot be two, (or the hunt would choose $x_1 = (3, 2)$). If $s \geq 4$ then $x_1$ has spectral value of at least 6, contradicting $(*)$ as above. Hence $s = 3$.

If $r = 1$ then $s$ can be at most 4, for otherwise the spectral value of $x_1$ is at least 6 which gives a contradiction as before. If $s = 2$ then $K_{S_1} + A_1$ is numerically trivial. □

We will now classify the possible surfaces $S_0$ and in each case prove log uniruledness using §6.

**19.3 If $x_0$ is a chain singularity.**

**19.3.1 Lemma.** *If $x_0$ is a chain singularity, then there is one singularity of $T_1$ on $\Sigma_1$, $\Sigma_1$ meets $E_1$ at a smooth point, and $K_{T_1} + \Sigma_1$ is log terminal.*

*Proof.* Since there are only two singular points along $E_1$, $\Sigma_1$ must meet $E_1$ at a smooth point. The rest now follows from the fact that $A_1$ is smooth, see (11.4). □

By (19.3.1) there are two singular points $x = x_0$ and $y$ on $S = S_0$. $y$ is the unique singular point on $\Sigma = \Sigma_1$. We say that we are 'adding' the singularity $y$. Recall that $E = E_1$ is a $-k$-curve.

Let $v$ be the coefficient of the curve meeting $\Sigma$ at $y$. Let $e = e_0$. Then $K_S \cdot \overline{\Sigma} = -1 + e + v$. Thus $S$ is a log del Pezzo iff $e + v < 1$.

We note in all cases there is a map $j : \tilde{S} \longrightarrow \mathbb{P}^2$, a composition of smooth blow ups. To describe germs of curves on $S$ we will follow the notation of example (6.9).

**19.3.2 Case 1.** Fix a configuration in $\mathbb{P}^2$ of a conic $B$, a secant line $A$, a tangent line $C$. Let $C \cap B = c$, $A \cap C = t$, and $A \cap B = \{a, b\}$. Let $L_{wz}$ indicate the line in $\mathbb{P}^2$ from $w$ to $z$, and $M_p$ the tangent line to the conic at $p$.

**19.3.2.1 Case 1:** $(s, r) = (3, 2)$.

**19.3.2.1.1 Add** $(2, 3, 2)$. $y = (A_{k-1}, 3', 3, 2)$. $e_0 = \frac{12k-24}{12k-17}$ $v = \frac{7}{8k+5}$. $e(3') = \frac{7k}{8k+5}$. For $S$ to be del Pezzo, and this to be a step of the hunt we have $e + v < 1$ and $e \geq e(3')$. The possibilities are $k = 4, 5$. $j$ is given by blowing up 3 times at $c$ along $C$, and once more over $c$, at the point where the $-1$-curve meets an exceptional $-2$-curve, 4 times at $b$ along $B$, and $k$ times at $a$ along $A$. $x = (2, 2^C, k^A, 2, 2, 2)$. $y = (2, 3, 3^B, A_{k-1})$. Here for example $k^A$ indicates that the strict transform of the line $A$ on $\tilde{S}$ is a $-k$-curve. $\Delta(x) = 31, 43$. $\Delta(y) = 37, 45$. Let $W = L_{ac}$ (on $S$). $W$ is a $-1$-curve and has two smooth branches at $y$, they are $(A_{k-2}, 2', 3, 3, 2)$ and $(A_{k-1}, 3, 3, 2')$. For $k = 4$, $-K_S \cdot W = 3/37$, thus $K_S$ is log uniruled by (6.5). Let $Z = M_b$. $Z$ is a $-1$-curve, and has two smooth branches at $x$. They are $(2, 2', k, 2, 2, 2)$ and $(2, 2, k, 2, 2', 2)$. When $k = 5$ one computes $-K_S \cdot W = 1/45$, $-K_S \cdot Z = 1/43$. $S$ is log uniruled by (6.5).

**19.3.2.1.2 Add $A_3$.** $y = (A_{k-1}, 3, 2, 2)$. $e_0 = \frac{24k-42}{24k-31}$ $v = \frac{3}{4k+3}$. $k = 3, 4$. $j$ : Blow up 4 times at $C, c$, 4 times at $B, b$ then $k$ times (above $b$) along $A$. Let $Z = L_{ac}$. $-K_S \cdot Z = 9/41$. Log uniruled by (6.5). For $k = 4$ let $W = M_a$. $W$ has two branches at $x$. One checks that each branch has index 65 $=\Delta(x)$. $-K_S \cdot W = 5/65$, log uniruled by (6.5).

**19.3.2.1.3 Add $A_2$.** $y = (A_{k-1}, 3, 2)$. $e = \frac{32k-56}{32k-44}$ $v = \frac{2}{3k+2}$. $e + v < 1$ gives $k < 4$, $k = 2$ not a hunt step, so $k = 3$. $j$ : Blow up 4 times at $B, b$, 4 times at $C, c$ and then 3 times above $c$ along $A$. $\Delta(x) = 52$. Let $Z = L_{bc}$. $Z$ has two branches at $x$, one lt. One checks the non lt branch has index 52. $-K_S \cdot Z = 4/52$.

**19.3.2.2 Case 1** $(s, r) = (4, 1)$.

**19.3.2.2.1 Add** $(3, 2, 2)$. $y = (A_{k-1}, 4', 2, 2)$. $e = \frac{10k-21}{10k-13}$ $v = \frac{6}{7k+3}$. $e + v < 1$ gives $k < 9$. $e \geq e(4') = \frac{6k}{7k+3}$ gives $k \geq 5$. $k = 5, 6, 7, 8$. Corresponding indices are $\Delta(x) = 37, 47, 57, 67$. $\Delta(y) = 38, 45, 52, 59$. $j$ : Blow up 5 times at $B, b$, 3 times at $C, c$, $k$ times at $A, a$. Let $Z = M_b$. $-K_S \cdot Z = \frac{k-9}{10k-13}$. Let $W = L_{ac}$. $-K_S \cdot W = \frac{k-9}{7k+3}$. $Z$ has two branches at $x$, one lt, one smooth; $W$ two at $y$, one lt, one smooth. Apply (6.5) to $Z$ for $k \neq 7, 8$, and (6.5) to $Z + W$ for $k = 8$. For $k = 7$ let $H$ be the conic of (6.9.1), having $3^{rd}$ order contact with $B$ at $c$ and tangent to $A$ at $a$. $H$ has two branches at $y$, one singular, one smooth, and one checks that each has local index 52. $-K_S \cdot H = 4/52$, so we can apply (6.5). □

**19.3.2.2.2 Add $A_1$ point.**
$y = (A_{k-1}, 3)$. $e = \frac{35k-55}{35k-43}$ $v = \frac{1}{2k+1}$. $e + v < 1$ gives $k \leq 4$. $k = 3, 4$. $\Delta(x) = 62, 97$. $\Delta(y) = 7, 9$. $j$ : 5 times at $B, b$, 3 times at $C, c$, $k$ times at $A, t$. For $k = 3$ take $H$ smooth conic tangent to $A$ at $a$ and with $3^{rd}$ order contact with $B$ at $c$. $H$ has two branches at $x$, (each singular) each of index 62. $-K_S \cdot H = 21/62$. For $k = 4$ let $Z = L_{bc}$. Two branches at $x$, one lt, one smooth. $-K_S \cdot Z = 3/97$. □

**19.3.2.2.3 Add $A_4$ point.** $y = (A_{k-1}, 3, 2, 2, 2)$, $e = \frac{14k-22}{14k-13}$, $v = \frac{4}{5k+4}$. $e + v < 1$ gives $k \leq 7$, $k = 3, 4, 5, 6, 7$. $\Delta(x) = 29, 43, 57, 71, 85$. $\Delta(y) = 19, 24, 29, 34, 39$. $j$: 3 times at $C, c$, 5 times at $B, b$ then $k$ times over $b$ along $A$. $Z = L_{ac}$, $W = M_a$. Each has two branches at $x$, one of which is singular. $-K_S \cdot Z = \frac{2(8-k)}{14k-13}$. When $\Delta(x)$ is prime, the singular branch of $Z$ is non-Cartier by (4.12.2), and we can apply (6.5). One checks that the non lt branch of $Z$ has index 57 when $k = 5$ and the non lt branch of $W$ has index 85 when $k = 7$. Since $-K$ is not a generator we can apply (6.5). □

**§19.3.3 Case 2.** Here fix the configuration as in Case 1, but with $A$ the conic and $B$ the secant line.

**19.3.3.1** $(r, s) = (2, 3)$.

**19.3.3.1.1 Add** $(3,2,2)$ **point.** $y = (A_k, 4, 2, 2)$, $e = \frac{21k-32}{21k-23}$, $v = \frac{6}{7k+10}$. $e + v < 1$ gives $k \leq 3$. But for $k < 4$ this is not a hunt step (the $-4$-curve has higher coefficient). So this case does not occur.

**19.3.3.1.2 Add** $(3,2)$ **point.** $y = (A_k, 4, 2)$, $e = \frac{28k-44}{28k-33}$, $v = \frac{4}{5k+7}$. $e + v < 1$ gives $k \leq 3$. $k = 2$ not a hunt step so $k = 3$. $\Delta(x) = 51$, $\Delta(y) = 22$.$j$ : 4 Times at $B, b$, 4 times at $c$, to get exceptional locus over $c$, $(-2, -3, -1, -2)$, with $C$ a $-2$-curve, then blow up 4 times above $c$ along $A$. $x = (2, 2^C, 3^B, 3^A, 2, 2, 2)$. Let $W = M_b$. $W$ has two branches at $x$, $(2, 2', 3, 3, 2, 2, 2)$ and $(2, 3, 3, 2, 2', 2)$, each of index 51. $-K_S \cdot W = 5/51$. □

**19.3.3.1.3 Add** $A_3$ **point.** $y = (A_k, 3, 2, 2)$, $e = \frac{35k-41}{35k-29}$, $v = \frac{4}{5k+7}$. $e + v < 1$ gives $k \leq 2$. $k = 2$ not a hunt step. So this case does not occur. □

**19.3.3.2 Case 2,** $(r, s) = (1, 4)$.

**19.3.3.2.1 Add** $(4,2)$ **point.** $y = (A_k, 5, 2)$, $e = \frac{15k-30}{15k-22}$, $v = \frac{6}{7k+9}$. $e + v < 1$ gives $k \leq 5$. $e \geq e(5) = \frac{6k+6}{7k+9}$ gives $k = 5$. $\Delta(x) = 53$. $j$ : 6 times at $A, a$,3 times at $C, c$, 5 times at $B, b$. $Z = L_{bc}$. $Z$ has two branches at $x$, one lt, one smooth. $-K_S \cdot Z = 2/53$. □

**19.3.3.2.2 Add** $A_2$ **point.** $y = (A_k, 3, 2)$, $e = \frac{35k-50}{35k-28}$, $v = \frac{2}{3k+5}$. $e + v < 1$ gives $k \leq 3$. $k = 2$ not hunt step so $k = 3$. $\Delta(x) = 67$. $j$: 5 times at $B, b$, 3 times at $C, c$ then 4 times at $A, c$. $x = (2^C, 4^B, 3^A, A_4)$. $W = M_b$. Two branches at $x$, one lt, one smooth. $-K_S \cdot W = 8/67$. □

**19.3.3.2.3 Add** $A_4$ **point.** $y = (A_k, 3, A_3)$, $e = \frac{21k-30}{21k-21}$, $v = \frac{4}{5k+9}$. $e + v < 1$ gives $k \leq 4$. $k = 3$ not a hunt step so $k = 4$. $\Delta(x) = 64$. $j$: 3 times at $C, c$, 5 times at $B, b$ then 5 at $A, b$. $W = L_{ac}$. $W$ has two branches at $x$, on lt, one singular. The singular branch has index 64. $-K_S \cdot W = 4/64$. □

**19.3.3.3 Case 2** $(r, s) = (1, 3)$.

**19.3.3.3.1 Add** $(3,2)$ **point.** $y = (A_k, 4, 2)$, $e = \frac{12k-24}{12k-17}$, $v = \frac{4}{5k+7}$. $e + v < 1$ gives $k \leq 8$. $e \geq e(4) = \frac{4k+4}{5k+7}$ gives $k = 4, 5, 6, 7, 8$. $\Delta(x) = 31, 43, 55, 67, 79$. $j$: 3 times at $C, c$, $k+1$ times at $A, a$, 4 times at $B, b$. $W = L_{bc}$. $W$ has two branches at $x$, on lt, one smooth. $-K_S \cdot W = \frac{9-k}{12k-17}$. One checks when $k = 6$, the smooth branch of $W$ has index 55. Thus (6.5) applies when $k \neq 8$. For $k = 8$ let $Z = M_a$. $Z$ has two branches at $y$, one lt, other smooth. $\Delta(y) = 47$ (so prime). $-K_S \cdot W = 1/47$. Now apply (6.5). □

**19.3.3.3.2 Add** $A_3$ **point.** $y = (A_k, 3, 2, 2)$, $e = \frac{15k-24}{15k-16}$, $v = \frac{3}{4k+7}$. $e + v < 1$ gives $k = 3, 4, 5, 6, 7$. $\Delta(x) = 29, 44, 59, 74, 89$. $j$: 3 times at $C, c$, 4 times at $B, b$ then $k+1$ at $A, b$. $Z = L_{ac}$. $Z$ has two branches at $x$, one lt, one singular. The singular is not Cartier by (4.12). $-K_S \cdot Z = \frac{2(8-k)}{15k-16}$. One checks for $k = 4, 6$ that the singular branch has index $44, 74$. □

**19.3.3.3.3 Add $A_2$ point.** $y = (A_k, 3, 2)$, $e = \frac{21k-32}{21k-23}$, $v = \frac{2}{3k+5}$. $e + v < 1$ gives $k = 3, 4, 5, 6$. $\Delta(x) = 37, 57, 77, 97$. $j$: 4 at $B, b$, 3 at $C, c$ then $k+1$ at $A, c$. $Z = M_b$. $Z$ has two branches at $x$, on lt, one singular. The singular branch is $(2^C, 3^B, k', 2', 2, 2)$. By (4.12) the singular branch is not Cartier. For $k = 4$ on checks the index of the singular branch is 37, 57. $-K_S \cdot Z = \frac{3(7-k)}{21k-23}$. Thus (6.5) applies for $k \neq 5$. For $k = 5$ let $W = M_a$. $W$ has one lt branch, one singular branch, at $x$. One checks the singular branch has index 77. Since $K_S$ is not a generator, (6.5) applies. □

**§19.4 If $x_0$ is a non-chain singularity.**

Thus there are three singularities along $E$, and along $A_1$. If $\Sigma$ meets $E$ at a smooth point, then $\pi(E)$ is smooth. It follows (since $A_1$ is smooth), that $\Sigma$ is in the Du Val locus, thus $a = 1$, a contradiction.

So we can assume $E$ meets $\Sigma$ at a singular point $z$. Now the possibilities for $\tilde{T} \longrightarrow \tilde{S}_1$ are given in (11.4). Since by (4.12.3) there must be a non-Du Val singularity along $\Sigma \setminus \Sigma \cap E$, it follows (in the language of (11.4)) that there are no interior blow ups, and $K_T + E + \Sigma$ is lt. Let $w$ be the other singularity along $\Sigma$.

**19.4.1.1 Case 1, $(s-1, r+1) = (2, 3)$.** By the classification of lt singularities, necessarily the singularities along $E$ are $A_3$, $A_2$ and $z = A_1$. Thus (since $x_0$ is not Du Val ), $k \geq 3$. By (11.4), $w = (3', A_{k-2}, 3, 3, 2)$. $e_0 = \frac{12k-24}{12k-23}$. One easily checks that $K_{S_0} \cdot \Sigma > 0$, a contradiction. □

**19.4.1.2 Case 1, $(s-1, r+1) = (3, 2)$.** The singularities along $A_1$ are $(3, 2, 2)$, $A_4$, and $A_1$.

Suppose first that $(3, 2, 2)$ is not added. Then necessarily the singularities along $E$ are $(3, 2, 2)$,$A_1$, and $z = A_1$, and $w = (3', A_{k-2}, 3, 2, 2, 2)$, $k \geq 2$. One checks that if $k \geq 3$ then $K_{S_0}$ is ample. Thus $k = 2$: The singularities are a non-chain $x$ with center $(-2)$ and branches $(3', 2, 2)$, $(2)$, $(2)$, and $y = w = (3, 3, 2, 2, 2)$. The map $\tilde{S}_0 \longrightarrow \mathbb{P}^2$ is obtained from a configuration with conic $B$, secant line $A$ and tangent line $C$. Blow up 3 times at $c$ along $C$, 5 times at $b$ along $B$ and then 3 more times over $b$ to give the prescribed configuration. $A$ becomes $E$, $C$ is a $-2$-curve, one of the branches, and $B$ is the $-3$-curve of the $(3, 2, 2)$ branch. Let $W$ be the tangent line at $a$. $W$ is a one curve on $S_0$ with two branches at $x$. One meets $C$, and has local index (one computes) 16, the other is tangent to $B$ and transversal to $A$ (at $a$). Its local index is 8. One computes

$$-K_{S_0} \cdot W = 3 - 3/8 - 2 \cdot 3/4 - 3/4 = 3/8.$$

Thus $S_0$ is log uniruled by (6.5).

Now we assume $(3, 2, 2)$ is added. Then $z$ has index either 2 or 3. Suppose first it is 2. Then $k \geq 3$, and $y = (3', A_{k-2}, 4, 2, 2)$. $e_0 = \frac{5k-10}{5k-9}$. One checks $K_{S_0}$ is ample.

Suppose $z = (2, 2)$. Then $k \geq 3$, $w = (4', A_{k-2}, 4, 2, 2)$, and $e_0 = \frac{30k-60}{30k-59}$. One checks $K_{S_0}$ is ample.

Finally suppose $z = (3)$. Then $k \geq 2$, $w = (2', 3, A_{k-2}, 4, 2, 2)$, and the coefficient of the $-3$-curve at $x_0$ is $\frac{21k-33}{30k-49}$. The coefficient of $2'$ is $\frac{7k+2}{21k+2}$. One checks $K_{S_0}$ is nef. □

**19.4.2.1 Case 2:** $(r, s) = (2, 3)$. Singularities along $A_1$ are $(3, 2, 2), A_3, (3, 2)$. This case is ruled out by index consideration. □

**19.4.2.2 Case 2:** $(r, s) = (1, 4)$. Singularities along $A_1$ are $(4', 2), A_4, (2, 2)$. Necessarily $(4, 2)$ is added, $z = (2)$, $k \geq 3$, and $y = (3', A_{k-1}, 5, 2)$. $e_0 = \frac{30k-60}{30k-59}$. One checks $K_{S_0}$ is ample. □

**19.4.2.3 Case 2:** $(r, s) = (1, 3)$. Singularities along $A_1$ are $(3', 2), A_3, (2, 2)$. In any case $z = (2)$.

Suppose first that $(3, 2)$ is added. Then $k \geq 3$ and $y = (3', A_{k-1}, 4, 2)$. $e_0 = \frac{12k-24}{12k-23}$. One checks $K_{S_0}$ is ample. □

## §20 Proof of (1.1) and corollaries

We begin with the formal proof of (1.1).

**20.1 Theorem.** *Let $S$ be a rank one log del Pezzo surface. Assume $S^0$ has trivial algebraic fundamental group. If $S$ does not have a tiger then $S^0$ is uniruled. Moreover, exactly one of the following holds for the hunt with $(S_0, \Delta_0) = (S, \emptyset)$:*

(1) *$A_1 \subset S_1$ is smooth, and $(S_2, A_2 + B_2)$ is a smooth Banana and included in the classification (13.2).*
(2) *$g(A_1) = 2$ and $S$ is the surface given in (15.2).*
(3) *$g(A_1) = 1$ and $A_1$ has a simple node. $(S_2, A_2 + B_2)$ is a fence with $B_2$ smooth. Either $(S_2, A_2 + B_2)$ is included in the classification (13.5), or we have (17.3.2).*

*Proof.* The possibilities for the first two hunt steps are described in (8.4.7). By (14.1), $T_1$ cannot be a net. Now apply (15.1), (16.1), (17.1), (17.1.1) and (18.1). □

*20.2 Proof of (1.1).* Notation as above (1.1). As above (1.3), the $(K_X + D)$-MMP reduces (1.1) to (1.3). By (6.2) and (7.2), we may assume $D$ is empty, $S$ has no tiger, and $S$ has trivial algebraic fundamental group. Now apply (20.1). □

Here we give proofs of the corollaries in the introduction, in those cases where the result does not follow immediately from (1.1) and (1.3).

*Proof of (1.4).* (3) and (4) are instances of (6.1). Case (2) reduces to case (1) after passing to a log terminal model, as in the proof of (6.3). Thus its enough to consider (1). Now pass to a log resolution, replace $B$ by the total transform, and apply (1.1). □

*Proof of (1.6).* Let $B = \llcorner \Delta \lrcorner$. If $S$ has Picard number one the result follows from (7.5) and (7.10). Otherwise, running $(K_S + \Delta)$-MMP we may assume $S$ has Picard number two, and each edge gives a $\mathbb{P}^1$-fibration, and $B$ has degree at most one for either fibration. As in the proof of (7.10), we can deform the union of the two general fibres $F_1 + F_2$ to get a dominating family of integral rational curves $D \subset S^0$ which deform with a general point fixed. $D \cdot B \leq 2$, so $D \cap U$ gives a connecting family of images of $\mathbb{A}^1_*$. □

*Proof of (1.7).* This follows from (3.3-3.4) of [5]. Batyrev works with terminal three folds, but his arguments apply without change to log surfaces. The description of the edges follows from his definition of total pullback, and our (1.3). □

*Proof of (1.8), due to Mori.* By (2.1) and (3.1) of [29] it is enough to show that if

$$T_S(-\log(D)) \longrightarrow L$$

is a generically surjective map to a line bundle $L$, then $c_1(L) \cdot H \geq 0$, for any ample $H$. By (1.7) it is enough to show $c_1(L) \cdot [C] \geq 0$, for $C$ as in (1.7). This follows from (5.5.1). □

*Proof of (1.8.1).* Immediate from (1.8), and (10.8), (10.12) of [27]. □

*Proof of (1.12).* We can assume $C$ is not rational.

Suppose first $\dim(X) = 2$. Since $K_X \cdot C < 0$, we have $C^2 > 0$ by adjunction. Thus the Kodaira dimension of $K_X + D$ is negative, and so by (1.1) $X$ is dominated by rational curves meeting $D$ at most once. The dominating family meets $C$ by the Hodge Index Theorem, and has no basepoints along $C$ by (5.5).

Now suppose $\dim(X) = 3$. Let $W \in |m(K_X+D)|$ for some $m > 0$. Let $\tilde{C}$ be the normalisation of $C$. As in the proof of (2.13) of [23], there is a proper smooth map $f : T \longrightarrow B$ from a smooth surface to a smooth curve, with $\tilde{C}$ a fibre of $f$, and a map $g : T \longrightarrow X$ with two dimensional image, such that $g|_{\tilde{C}}$ is the normalisation map (in short $C$ sweeps out a surface). Let $S = \overline{g(T)}$. Since $S$ is covered by $W$ negative curves, $S$ is a component of $W$, and $S \cdot C < 0$. Thus $K_S \cdot C < 0$.

Let $p : S' \longrightarrow S$ be the normalisation. $g$ lifts to $g' : T \longrightarrow S'$. Let $C' = g'(\tilde{C})$, $\pi : S'' \longrightarrow S'$ be the minimal desingularisation, $C'' \subset S''$ be the strict transform of $C'$, $V = D|S$, $V' = p^*(V)$, and $V'' = \pi^*(V')$.

There is an effective integral Weil divisor $\Gamma$ such that $K_{S'} + \Gamma = p^*(K_S)$. Recall that the intersection product with $\mathbb{Q}$-coefficients is defined on any normal surface. We have

$$\begin{aligned}(K_{S''} + V'') \cdot C'' &\leq (K_{S'} + V') \cdot C' \\ &\leq (K_{S'} + \Gamma + V') \cdot C' = (K_S + V) \cdot C < 0\end{aligned}$$

where the third inequality holds since $C'$ sweeps out $S'$.

Thus by the surface case there is a rational curve through a general point of $C''$, meeting $V''$ at most once. Of course its image passes through a general point of $C$ and meets $D$ at most once. □

## §21 A SURFACE WITH $\pi_1^{\mathrm{alg}}(S^0) = \{1\}$ BUT NO TIGER

Our first goal is to give an example of $S$ without tiger. We again use the hunt, but of course, now we cannot assume $S = S_0$ does not have a tiger. None the less, following (8.2), (8.2.5), and (8.2.8), we construct $(S_i, \Delta)$ starting with $(S_0, \Delta_0) = (S, \emptyset)$, by exactly the same procedure used in hunting for a tiger. We continue to use the notation (8.2.10).

By (8.2.5.8), if $K_{S_{i+1}+\Delta_{i+1}}$ has a tiger, so does $K_S$. To get an implication in the other direction, we need to look at a smaller boundary.

**21.1 The boundaries $\gamma_i$.** We introduce some more notation. We recall our convention of using the same symbol for a curve, and its strict transform under some birational map. Inductively define

$\gamma_i = \sum_{j\leq i} m_j E_j$ and $m_{i+1} = e(E_{i+1}, K_{S_i} + \gamma_i)$. We let $\gamma_0 = \emptyset$. Thus $\gamma_1 = e_0 A_1$, $m_2 = e(E_2, K_{S_1} + e_0 A_1)$, and $\gamma_2 = e_0 A + m_2 B$.

For the next lemma, assume $T_j$ is not a net for $j \leq i$.

**21.1.1 Lemma.** *If $K_S$ has a tiger, then so does $K_{S_i} + \gamma_i$.*

*Proof.* We prove inductively that if $K_{S_i} + \gamma_i$ has a tiger, then so does $K_{S_{i+1}} + \gamma_{i+1}$.

Suppose $K_{S_i} + \gamma_i + \alpha$ is numerically trivial, where $\alpha$ is effective. Define effective $\beta$ by

$$\begin{aligned} f_i^*(K_{S_i} + \gamma_i + \alpha) &= K_{T_{i+1}} + \gamma_i + m_{i+1}E_{i+1} + \beta \\ &= \pi_{i+1}^*(K_{S_{i+1}} + \gamma_{i+1} + \pi_{i+1}(\beta)). \end{aligned}$$

Then $K_{S_i} + \gamma_i + \alpha$ is klt iff $K_{S_{i+1}} + \gamma_{i+1} + \pi_{i+1}(\beta)$ is klt. □

*21.1.2.* We note that (21.1.1), and its proof, hold for any sequence of relative Picard number one extractions and contractions $f_i : T_{i+1} \longrightarrow S_i$ and $\pi_{i+1} : T_{i+1} \longrightarrow S_{i+1}$, with $f_i$ $K$-positive. Indeed, all we use in the proof is that $\beta$ is effective.

**21.2 A condition to ensure there is no tiger.**

Let $\Gamma$ be an effective $\mathbb{Q}$-Weil divisor. For a point $p$ on a smooth surface, let $m_p(\Gamma)$ be the coefficient of the exceptional divisor of the blow up of $p$ in the inverse image of $\Gamma$ (when $\Gamma$ is reduced this is the usual multiplicity).

**21.2.1 Lemma.** *Let $X$ and $Y$ be divisors meeting normally at a smooth point $p$ of a surface $S$. If*

$$\max(a, b) + m_p(G) < 1$$

*then $K_S + aX + bY + G$ is klt at $p$.*

*Proof.* We may assume $a \geq b$. Blow up $S$ at $p$. The exceptional divisor $E$ has coefficient $a + b + m_p - 1 < b$, and the strict transform of $X$ and $Y$ union $E$ has normal crossings. Since $m_p$ does not go up under blow ups, the result follows by induction. □

**21.2.2 Proposition.** *Let $B$, $A$, $C$ be a conic, a secant line and a tangent line in $\mathbb{P}^2$, with $A \cap B \cap C = \emptyset$. $K_{\mathbb{P}^2} + (60/67)A + (381/469)B + (30/67)C$ has no tiger.*

*Proof.* Let $\alpha$ be an effective $\mathbb{Q}$-Weil divisor such that

$$K + \Delta = K_{\mathbb{P}^2} + (60/67)A + (381/469)B + (30/67) + \alpha$$

is numerically trivial.

Let $p$ be a point of $\mathbb{P}^2$, and let $L$ be a line. Note that $m_p(\alpha) \leq \alpha \cdot L = 3 - 60/67 - 2.381/469 - 30/67 = 15/469$. Let $\Sigma^i$ be the $-1$-curve of the $i^{th}$ blow up of $\mathbb{P}^2$ at $z = B \cap C$ along $B$. Then

$$e(\Sigma^1, K + \Delta) \leq 137/469.$$
$$e(\Sigma^2, K + \Delta) \leq 274/469.$$

Now blow up twice at $z$ along $B$. Applying (21.2.1), to the total transform, which has normal crossings, we conclude that $K + \Delta$ is klt. $\square$

*Remark.* The coefficients in (21.2.2) may seem a bit bizarre. We chose them to make the analysis in the next section as simple as possible.

**21.3 An example of a smooth Banana with no tiger.**

Throughout (21.3) let $S = S_0$.

**21.3.1 Lemma.** *If $S_2$ is a smooth banana, then $\pi_1(S^0)$ is Abelian.*

*Proof.* By (13.2), after a blow up and blow down of $S_2$ we obtain $\overline{\mathbb{F}}_2$. We will use the hunt notation (although it is possible this last transformation is not part of the hunt). We have by (13.2) a rational map $h : S \dashrightarrow \overline{\mathbb{F}}_2$, and the exceptional locus of $h^{-1}$ is contained in $A + B + C \subset \overline{\mathbb{F}}_2$, for $A$, $B$ two sections and $C$ a fibre, such that $A + B + C$ has normal crossings. Blowing up a point of $p \in A \cap B$ and blowing down the fibre through $p$, gives a rational map $g : S \dashrightarrow \mathbb{P}^2$ such that the exceptional locus of $g^{-1}$ is contained in $W \subset \mathbb{P}^2$, a union of four lines, with normal crossings. Then $U = \mathbb{P}^2 \setminus W$ is an open subset of $S^0$, thus $\pi_1(U) \twoheadrightarrow \pi_1(S^0)$. $\pi_1(U)$ is Abelian by Zariski's conjecture. $\square$

**21.3.2 Lemma.** *If $S_2$ is a Banana of type given in (19.3.2.2.1), then $S^0$ is simply connected.*

*Proof.* By (21.3.1), $\pi_1(S^0) = H_1(S^0, \mathbb{Z})$. If $D$ is the exceptional locus of $\tilde{S} \longrightarrow S$, then according to (2.1) of [33] there is an exact sequence

$$0 \longrightarrow r : \mathrm{Pic}(\tilde{S}) \longrightarrow H^2(D, \mathbb{Z}) \longrightarrow H_1(S^0, \mathbb{Z}) \longrightarrow 0$$

Let $M$ be the cokernel.

If we let $\mathcal{D} \subset \operatorname{Pic}(\tilde{S})$ be the subgroup generated by $D$, then (since $x$ and $y$ are chain singularities), we have

$$0 \longrightarrow \mathcal{D} \longrightarrow H^2(D,\mathbb{Z}) \longrightarrow \frac{\mathbb{Z}}{\Delta(x)\cdot\mathbb{Z}} \oplus \frac{\mathbb{Z}}{\Delta(y)\cdot\mathbb{Z}} \longrightarrow 0$$

with either end component mapping to the generator in $\frac{\mathbb{Z}}{\Delta(p)\cdot\mathbb{Z}}$ (for $p = x, y$).

Let $E_c$, $E_b$ be the $-1$-curve over $c$, $b$ (in the notation of (19.3.2)). One easily checks that the image of $E_b$ in $M$ is $(1,3)$ and the image of $E_c$ in $M$ is $(1,2)$. Thus $r$ is surjective. □

**21.3.3 Example.** *If $S$ is the surface from (19.3.2.2.1) with $k = 8$ then $S^0$ is simply connected, and $S$ has no tiger.*

*Proof.* Let $f_1$ and $f_2$ be (as usual) the first two hunt steps. $(S_2, A + B)$ is a smooth banana. Let $f_3$ blow up the $A_1$ point along $A \subset S_2$ (this is not a hunt step), so $(S_3, B + A + C)$ is as in (21.2.2). $m_1 = e_0 = \frac{60}{67}$. $x_1$ has the form (with respect to $A_1$) $(\underline{3}, 2, 2)$. Thus by (8.3.8), $m_2 = 3/7 + 3e_0/7 = 381/469$, and $m_3 = (1/2)(60/67) = 30/67$. By (21.2.2), $K_{S_3} + \gamma_3$ has no tiger. We conclude $S$ has no tiger by (21.1.1). $S^0$ is simply connected by (21.3.2). □

### 21.4 Counter-Example to a conjecture of Miyanishi.

**21.4.1 Definition.** We say that a surface $V$ is **affine ruled** if there is an open subset of $V$ which is isomorphic to a product $U \times \mathbb{A}^1$.

According to [15], Miyanishi has conjectured that given any rank one log del Pezzo $S$ there is a finite étale cover of $S^0$ which is affine ruled. We show here that any tiger-free $S$ with $S^0$ simply connected (for example the surface of (21.3.3) ) is a counter-example.

**21.4.2 Lemma.** *Let $S$ be a rank one log del Pezzo, and $V \subset S^0$ an open subset dominated by images of $\mathbb{A}^1$. If the complement of $V$ contains a divisor, then $S$ has a tiger.*

*Proof.* Assume that there is a dominant map $f : U \times \mathbb{A}^1 \longrightarrow V$, and that $D \subset V^c \subset S$ is an irreducible curve. Let $C$ be the closure of the image of a general $\mathbb{A}^1$. If $C$ meets $D$ at a general point of $D$, then $(K_S + D) \cdot C < 0$ by (5.11), and thus $D$ is a tiger.

Otherwise every $C$ passes through a common point $p = C \cap D$. Let $S' \longrightarrow S$ extract the divisor $E$ over $p$, so that $C'$ cuts out $E$. Then $C'$ and $D'$ are disjoint on $T$ (here the prime indicates strict transform). Define $\lambda$ so that $K_T + E + \lambda D' = f^*(K_S + \lambda D)$. Note $\lambda > 0$ since $S$ is lt. Now $C'$ meets $E$ in a single smooth point, and so by (5.11), $(K_T + E) \cdot C' = (K_S + \lambda D) \cdot C < 0$. Thus $K_S + \lambda D$ is anti-ample. It is not klt by construction, and so $\lambda D$ is a special tiger for $K_S$. □

**21.4.3 Corollary.** *An affine ruled rank one log del Pezzo surface has a tiger.*

**21.4.4.** *Let $S$ be a rank one log del Pezzo. If $S$ does not have a tiger and $\pi_1^{\mathrm{alg}}(S^0) = \{1\}$, then no open subset of $S$ has a finite étale cover which is affine ruled.*

*Proof.* Let $S$ be a simply connected rank one log del Pezzo, with no tiger. Suppose $V \subset S$ has a finite étale cover which is affine ruled. Dropping points from $V$, we may assume $V \subset S^0$. By (21.4.2) $V^c$ has codimension at least two, and thus $V$ is simply connected (see the proof of (7.3)). Thus $S$ is affine ruled, contradicting (21.4.3). □

*Remark.* Definition (21.4.1) is due to Miyanishi. Kollár has suggested that in view of the definition of ruled, (see for example (IV.1.1) of [26]) it is perhaps more natural to assume only the existence of a birational map $f : U \times \mathbb{A}^1 \longrightarrow S^0$. We do not know if (21.4.3) holds under this weaker assumption.

## §22 Tigers, complements and toric pairs

The following section prepares the way for the classification that appears in §23. First we introduce the notion of complements and give the connection between tigers and complements.

We first recall the definition of a complement, cf. Chapter 19 of [27], for more details.

**22.1 Definition.** Let $X$ be a normal variety and $\Delta = D + B$ a boundary, where $D$ is reduced and $B$ is a pure boundary. We say $K_X + \Delta$ is $n$-complemented, for a positive integer $n$, if there is a divisor $D'$, such that

(1) $nD' \in |-nK_X|$,
(2) $nD' \geq -nD - \llcorner (n+1)B \lrcorner$ and
(3) $K_X + D'$ is lc.

On a first encounter, (22.1) may seem a bit bizarre. For our classification in §23, we need only a simple form. For the readers convenience, we restate the definition in this case:

**22.1.4 Definition (1-complement of a reduced divisor).** $(X, D)$ as in (22.1). We say $K_X + D$ is 1-complemented if there is a reduced divisor $Y$ such that $K_X + D + Y$ is Cartier, trivial, and log canonical. In the notation of (22.1), $D' = D + Y$.

The following result, and its proof, are due to Shokurov. We learned of them from Alessio Corti:

**22.2 Lemma.** *Let $S$ be a Fano surface of Picard number one, and $\Delta$ an effective $\mathbb{Q}$-divisor, such that $K_S + \Delta$ is log canonical. If $K_S + \Delta$ has a tiger, then $K_S + \Delta$ has a 1, 2, 3, 4, or 6-complement.*

*Proof.* After scaling and applying (19.2) of [27] we may assume $K_S + \Delta$ is anti-nef and maximally log canonical. There is an extraction $\pi : T \longrightarrow S$ such that $K_T + \Gamma = \pi^*(K_S + \Delta)$ is lc, and

$D = \llcorner\Gamma\lrcorner$ is non-empty, and contained in the smooth locus. By (19.2) of [27] it is enough to show $K_T + \Gamma = K_T + D + B$ has a 1, 2, 3, 4 or 6-complement.

Look at the natural restriction exact sequence

$$0 \longrightarrow \mathcal{O}_T(-nK_T - (n+1)D - \llcorner(n+1)B\lrcorner) \longrightarrow \mathcal{O}_T(-nK_T - nD - \llcorner(n+1)B\lrcorner) \longrightarrow$$
$$\longrightarrow \mathcal{O}_D(-nK_D - \llcorner(n+1)B|_D\lrcorner) \longrightarrow 0.$$

Now $K_D + B|_D$ is $n$-complemented, for some $n \in \{1, 2, 3, 4, 6\}$ by (19.4) of [27]. We argue next that

$$H^1(T, \mathcal{O}_T(-nK_T - (n+1)D - \llcorner(n+1)B\lrcorner)) = H^1(T, K_T + \ulcorner-(n+1)(K_T + \Gamma)\urcorner) = 0.$$

If $K_S + \Delta$ is anti-ample, this holds by Kawamata-Viehweg vanishing [21], so we may assume $K_S + \Delta$ is numerically trivial. Let $G = K_T + \ulcorner-(n+1)(K_T + \Gamma)\urcorner$. $R^1\pi_*(G) = 0$ by Kawamata-Viehweg vanishing. Hence we are reduced to showing $H^1(S, \pi_*(G)) = 0$.

Now $\pi_*(G) = \mathcal{O}_S(K_S + \ulcorner-(n+1)(K_S + \Delta)\urcorner)$. Since the Picard number of $S$ is one, $\ulcorner-(n+1)(K_S + \Delta)\urcorner$ is ample, and Kawamata-Viehweg vanishing applies, unless $-(n+1)(K_S + \Delta)$ is integral, in which case (22.2.1) applies.

Thus we may find a divisor $D'$ on $T$, such that $nD' \in |-nK_T|$ and $K_T + D'$ is lc in a neighbourhood of $D$. Now the divisor $K_T + (1-\epsilon)D' + \epsilon(D + B)$ is not klt in a neighbourhood of $D$. By the connectedness Theorem, (17.4) of [27], it follows that this divisor is klt away from $D$. Thus $K_T + D'$ is lc and $K_T + \Gamma$ has a complement. □

**22.2.1 Lemma.** *Let $X$ be a $\mathbb{Q}$-factorial log Fano variety and $H$ a numerically trivial integral Weil divisor on $X$. Then $H^i(X, \mathcal{O}_X(K_X + H)) = 0$ for all $i > 0$.*

*Proof.* $H$ induces a cyclic Galois cover, étale in codimension one, $g : Y \longrightarrow X$ such that $K_Y = g^*(K_X + H)$. In particular, $Y$ is log Fano, and so $H^i(\mathcal{O}_Y(K_Y)) = 0$. $\mathcal{O}_X(K_X + H)$ is a direct summand of $g_*(\mathcal{O}_Y(K_Y))$. □

The following Lemma, refines (22.2), in a special case. All we need for this paper is the case (22.3.1), which will be used in (23.7).

**22.3 Lemma.** *Let $S$ be a log del Pezzo of rank one. Suppose $K_S + A$ is plt and $-(K_S + A)$ is ample, where $A$ is integral and non-empty. $A$ contains at most three singularities of $S$. Let $p \geq q \geq r$ be the indices of these singularities (we allow index 1).*

*Then $K_S + A$ has an $n$-complement $X$. Moreover*

(1) *$n = 1$, and $K_S + C$ is plt for any component $C$ of $X$, in the case $r = 1$ (that is when there are at most two singularities along $A$).*

*Otherwise $A$ has three singularities, $X$ has one other component $B$, $K_S + A + B$ is lc in a neighbourhood of $A$, and $A$ meets $B$ at one point. Moreover*

(2) $n = 2$ *and* $X = A + B$, *when* $q = r = 2$.

*Otherwise* $X = A + (B/n)$. *Moreover*

(3) $n = 3$, *when* $p = q = 3$.
(4) $n = 4$, *when* $p = 4$ *and* $q = 3$,
(5) $n = 6$, *when* $p = 5$ *and* $q = 3$.

*Proof.* $A$ can contain at most three singularities by adjunction. By the proof of (22.2), we know that if $K_A + \mathrm{Diff}_A(\emptyset)$ (see (L.2)) has an $n$-complement, then so does $K_S + A$. Moreover since we are lifting an element of $|-nK_A|$, we have a local picture around $A$. Now $A$ is a copy of $\mathbb{P}^1$, and the description of $X$ follows from the division into cases in the proof of (19.4) of [27]. Note in particular that if $R = X - A$, then $R \cdot A = 1/p$ when $n = 2$ and $R \cdot A = 1/sn$ when $n \geq 3$, where $s$ is the index of the point where $A$ meets $B$. On the other hand $B \cdot A \geq 1/s$, with equality in the case that $K_S + A + B$ is lc in a neighbourhood of $A$ and $R = (k/n)B$, for some $k \geq 1$. The result now follows, except in the case $n = 2$. Here we make the ad hoc argument that we have a classification cf. (23.6) and that $B$ is the image of the obvious fibre. □

**Toric Pairs.**

We now recall the definition of a toric variety. As we shall see in §23, they are fundamental to the classification of all log del Pezzos. Also we present a result about toric varieties which is rather surprising, in light of the boundedness results of Alexeev [2].

**22.4 Definition.** A pair $(S, X)$ is said to be toric, if $S \setminus X$ is a copy of a torus, and the natural action of the torus on itself, extends to an action on $S$.

Toric pairs are very easily understood. Either one may use the approach of fans, in which case a pair $(S, X)$ is essentially specified by two pairs of relatively prime integers (see for example [10]) or one may untwist a toric pair, using a series of transformations of type (a) and thus create $(S, X)$ inductively, see (23.9) and the proof of (23.12). The results below illustrate this idea.

Several people have asked the following question: Does the set

$$\{\, K_S^2 \mid S \text{ is a log del Pezzo surface of rank one} \,\}$$

satisfy ACC for bounded sequences ?

We prove

**22.5 Proposition.**

(1) *The set*

$$\{\, K_S^2 \,|\, S \text{ is a toric surface of rank one, with two singularities} \}$$

*is dense in the set of real numbers at least four, and*

(2) *The set*

$$\{\, K_S^2 \,|\, S \text{ is a toric surface of rank one} \}$$

*is dense in the set of positive real numbers.*

**22.6 Lemma.** *Let $S$ be a toric surface of Picard number one, with three singular points of index $p$, $q$ and $r$.*

*Then*

$$K_S^2 = \frac{(p+q+r)^2}{(pqr)}.$$

*Proof.* Now there are three toric divisors, $D_1$, $D_2$ and $D_3$, which form a triangle, such that each vertex is one of the singular points. Moreover $K_S + D_1 + D_2 + D_3$ is lc and linearly equivalent to zero. Thus

$$K_S^2 = (D_1 + D_2 + D_3)^2.$$

But as $K_S + D_1 + D_2 + D_3$ is lt,

$$\{\, D_i \cdot D_j \,|\, 1 \leq i < j \leq 3 \,\} = \{\, p, q, r \,\}.$$

The result is now an easy computation. □

**22.7 Lemma.**

(1) *The set*

$$\{ \frac{(x+1)^2}{x} \,|\, x \text{ is a positive real number} \}$$

*is dense in the set of real numbers at least four.*

(2) *Fix positive integers $N$ and $r$. The set*

$$I_{N,r} = \{ \frac{p'}{q} | \, c \text{ and } q > N \text{ are coprime integers, } p' = r(c+1) - q > 0 \,\}$$

*is dense in the set of positive real numbers.*

*Proof.* (1) is elementary calculus and (2) is clear. □

*Proof of Proposition.* Pick a positive integer $r$ and coprime integers $c$ and $q$. The three vectors $\{(1,0),(1,r),(-c,q)\}$ specify a fan in the lattice $\mathbb{Z}^2$, and give rise to a toric surface $S$ of Picard number one, with indices $p = rc - q$, $q$ and $r$.

By (22.6)

$$\begin{aligned} K_S^2 &= \frac{(p+q+r)^2}{(pqr)} \\ &= (1/r)\frac{((p/q)+1+(r/q))^2}{(p/q)} \\ &= (1/r)\frac{((p'/q)+1)^2}{(p'/q)}(1+\epsilon) \end{aligned}$$

where $|\epsilon| = o(1)$ (where we think of fixing $r$ and let $q$ go to infinity) and $p' = p + r$. The result now follows easily from (1) and (2) of (22.7). □

## §23 Classification of all but a bounded family of rank one log del Pezzo surfaces

Here we will give a classification of all but a bounded family of rank one log del Pezzos. The proofs use the methods of the hunt, introduced in §8, but as we expect there will be readers who are interested in the statement of the classification, independent of the rest of the paper, we will state the results in a self contained way. This will involve some repetition of definitions from §8. The analysis of this section is independent from most of the paper. It uses results and notations from §2-3 and §8-13 only.

Our classification is of the following sort. Starting with a rank one log del Pezzo we will make an essentially canonical birational transformation (the first hunt step) from $S$ to either a $\mathbb{P}^1$-fibration or to a second rank one log del Pezzo. We will explicitly classify the resulting surfaces. We will also indicate how the inverse transformations can be classified, the procedure is easy and elementary, but we do not write down an explicit list as it would be notationally too involved.

**23.0 The First Hunt Step.** Given a rank one log del Pezzo $S$, let $E$ be an exceptional divisor of the minimal desingularisation with maximal coefficient (that is minimal discrepancy). Let $f : T \longrightarrow S$ be the extraction (of relative Picard number one) of $E$. We assume $E$ is chosen so that $K_T + E$ is log terminal (such a choice is always possible, see (8.3)) $T$ has a unique $K_T$-negative contraction $\pi$. $\pi$ is either a $\mathbb{P}^1$-fibration, or birational. In the second case we let $S_1$ be the image, and let $A_1 \subset S_1$ be the image (with reduced structure) of $E$. $S_1$ is again a rank one log del Pezzo. For details see (8.3).

**23.1 Lemma.** *(Notation as in (23.0)) For all but a bounded family of rank one log del Pezzos, the Kodaira dimension of* $-(K_T+E)$ *is non-negative and either* $\pi$ *is a* $\mathbb{P}^1$*-fibration or* $K_{S_1}+A_1$ *is log canonical and anti-nef.*

**23.2 Theorem.** *(Notation as in (23.0)) Let* $S$ *be a rank one log del Pezzo surface, such that the Kodaira dimension* $-(K_T+E)$ *is non-negative. One of the following holds:*

(1) $\pi$ *is a* $\mathbb{P}^1$*-fibration, and* $-(K_T+E)$ *is nef,* $E$ *is a multi-section of degree at most two and* $-(K_T+E)$ *is ample if* $E$ *is a section.*
(2) $K_{S_1}+A_1$ *is lt and anti-ample and* $A_1$ *contains at most two singular points. Possibilities for* $(S_1, A_1)$ *are classified in (23.12).*
(3) $K_{S_1}+A_1$ *is lt and anti-ample and* $A_1$ *contains exactly three singularities. Pairs* $(S_1, A_1)$ *are classified in (23.5.1). They are in one to one correspondence with non cyclic quotient singularities (of dimension two)*

*Otherwise* $S_1$ *is Du Val, and one of the following holds:*

(4) $A_1 \subset S_1^0$ *is a rational curve of arithmetic genus one. Possibilities for* $S$ *such that* $A_1$ *has a cusp are bounded.*
(5) $K_{S_1}+A_1$ *is lt and numerically trivial, and* $A_1$ *contains exactly three singular points. Pairs* $(S_1, A_1)$ *fall into* 5 *families, listed in (23.5.2-4).*

**23.3 Lemma.** *Let* $(S_1, A_1)$ *be a pair of a rank one log del Pezzo with a reduced rational curve such that* $K_{S_1}+A_1$ *is anti-nef. If* $\pi : T \longrightarrow S_1$ *is a contraction of relative Picard number one, such that* $E$*, the strict transform of* $A_1$*, contracts* $f : T \longrightarrow S$ *to a log terminal surface, then* $S$ *is a rank one log del Pezzo.*

We now make remarks regarding the above:

*23.4 Remarks.*

(1) Rank one Du Val del Pezzo surfaces are bounded, and there is a short list of possibilities. See §3, or [33].
(2) It is a simple matter to classify possible $\mathbb{P}^1$-fibrations in (23.2.1), along the lines of (11.5). One can classify pairs in (23.2.4) by considering the anti-canonical series $|-K_{S_1}|$, see for example (3.6).
(3) Given a pair $(S_1, A_1)$ contractions $\pi : T \longrightarrow S_1$ such that $K_T+E$ is lt are classified in (11.4). Using the classification of log terminal singularities, one can then further classify possibilities such that $E$ contracts to a quotient singularity. In this sense (23.1-3) give a classification of all but a bounded family of log del Pezzo surfaces.
(4) Note the assumption in (23.2), is exactly that $E$ is a tiger. We believe it would be possible, using the methods of this paper, to completely classify those log del Pezzos $S$, such

that $E$ is not a tiger. Indeed, in the body of the paper we have already constructed an explicit collection $\mathfrak{F}$ which includes all $S$ with no tiger, under the additional assumption that $\pi_1^{\mathrm{alg}}(S^0) = \{1\}$.

**Proofs of (23.1-3).**

*Proof of (23.1).* By (9.3), it is enough to find $e' < 1$ such that if $e = e(S) > e'$, then $E_1$ is a tiger. Obviously we may assume $a$ (of the hunt) is less than one.

If $T$ is a net, and $e > 2/3$, then $E$ has degree at most 2, thus $K_T + E$ is anti-nef. Now suppose $\pi$ is birational. $K_{S_1} + aA_1$ is flush by (8.4.5). If $e > 4/5$ then $K_{S_1} + A_1$ is lc by (8.0.9.3) and, by (5.3) of [25], if $e > 41/42$, then $K_{S_1} + A_1$ is anti-nef. □

*Proof of (23.3).* Suppose $S$ is not a log del Pezzo. Then $K_S$ is nef, and pulling up and pushing down $K_{S_1} + eA_1$ is nef. Since $S$ is lt, $e < 1$, and we get a contradiction. Hence (2). □

*Proof of (23.2).* Clearly (1) holds if $T$ is a net. So assume $\pi$ is birational. We will show that $K_{S_1} + A_1$ is anti-nef, log terminal at singular points, and, except for a bounded collection of $S$, log canonical. The division into cases then follows easily by adjunction.

By assumption the Kodaira dimension of $-(K_{T_1} + E)$ is non-negative, thus $K_{S_1} + A_1$ is anti-nef. Suppose $a$ (from the hunt) is at least one. Then $K_{S_1} + E_1$ is $\pi$ non-positive, thus, since $K_{T_1} + E_1$ is log terminal, so is $K_{S_1} + A_1$.

So we may assume $a < 1$. $K_{S_1} + aA_1$ is flush by (8.4.5), thus $K_{S_1} + A_1$ is log terminal at singular points, by (8.0.4). If $e > 4/5$ then $K_{S_1} + A_1$ is lc by (8.0.9.3). But if $e < 4/5$ then $S$ is bounded by (9.3). □

**23.5 Proposition.** *Let $(S_1, A_1)$ be a pair of a rank one log del Pezzo surface, $S_1$ and a reduced irreducible curve $A_1 \subset S_1$ such that $K_{S_1} + A_1$ is anti-nef and log terminal. Assume there are at least three singularities of $S_1$ along $A_1$.*

*If $-(K_{S_1} + A_1)$ is ample then*

*(1) $S_1 \setminus A_1$ has exactly one singular point, a non-cyclic singularity, $z$. If $Z \longrightarrow S_1$ extracts the central exceptional divisor of $z$, then $Z$ is a $\mathbb{P}^1$-fibration and $E$ and $A$ are sections. $(S_1, A_1)$ is uniquely determined by $z$, and all non-cyclic singularities, $z$, occur in this way for some pair $(S_1, A_1)$.*

*If $K_{S_1} + A_1$ is numerically trivial then $S_1$ is Du Val, and $(S_1, A_1)$ is in one of five families:*

*(2) $S = S(A_1 + 2A_3)$. $(S_1, A_1)$ is given by (19.2), Case 1, with $s = 3$ and $r = 1$.*

*(3) $S = S(3A_2)$. $(S_1, A_1)$ is given by (19.2), Case 2, with $s = 2$ and $r = 1$.*

*(4) $K^2_{S_1} = 1$, $A_1$ is a $-1$-curve and $S_1$ is one of $S(A_1 + A_2 + A_5)$, $S(2A_1 + A_3)$, $S(4A_1)$. The pairs are obtained from (13.5.1), (13.5.3) and (13.5.6) respectively, by the inverse of a transformation of type (11.1.1).*

*Remarks.*

(1) For a construction, in (1), of $(S_1, A_1)$ from $z$, see the proof of (23.5.1.1)

(2) (23.5) is a classification of abstract pairs $(S_1, A_1)$, but our principal interest comes from (23.2), where the pair $(S_1, A_1)$ is the first hunt step, and this is why we use this subscripted notation. In fact, it is easy to check using (23.3) and (11.4) that each of the pairs $(S_1, A_1)$ actually occurs as the first hunt step for some $S$.

*Proof.* By (23.5.1.1) below, $z$ in case (1) determines $(S_1, A_1)$, and $K_{S_1} + A_1$ is negative. In (2-4) one can compute directly that $K_{S_1} + A_1$ is numerically trivial. We now show (without distinguishing between negative and numerically trivial) that any pair $(S_1, A_1)$ falls into one of the cases (1-4).

Let $r$ be the maximal index of any singularity along $A_1$.

Choose a singular point $y \in S_1$ as follows. If there is a point $y \notin A_1$ with $e(y) \geq \frac{r-1}{r}$, choose such a $y$ of maximal coefficient (for $K_{S_1}$). Otherwise take $y$ a point of $A_1$ of index $r$, which has maximal spectral value (among points of index $r$). One checks using (8.3.8), that $y$ is (one of the possible choices for) $x$ for the next hunt step with respect to $K_{S_1} + aA$ for any $1 - \delta <\geq a < 1$, for fixed $\delta > 0$ sufficiently small. (Note that some care is needed to define $x$, to ensure that it is independent of $a$.)

Now we run the hunt with respect to $K_{S_1} + aA$ for $a$ close to one, and $x = y$ as above. Note $K_{S_1} + aA$ is flush.

Here the situation is slightly different than in (8.4.7). We cannot assume that there is no tiger, indeed $A$ is obviously a tiger. None the less, the division into cases (8.4.7) will be the same, using essentially the same argument. This will require a bit of preliminary analysis.

To avoid having any coefficients larger than one, we alter slightly the scaling convention of (8.2.4). Define $b \geq e_1$ such that $K_T + aA_1 + bE$ is $R$-trivial ($R$ is the $K_T$-negative extremal ray). Let $\Gamma' = aA_1 + eE$.

**Claim.** *$b < a$, unless $T$ is a net, $K + A_1$ is negative and $x \notin A_1$, in which case $b = 2 - a$.*

*Proof.* If $K_{S_1} + A_1$ is numerically trivial, then $b = e_1 < a$, so we can assume $K + A_1$ is negative.

Now suppose $T$ is a net, and $x \notin A_1$. Since (for $a = 1$) $e_1 \geq 1/2$, $A$ and $E$ are sections, so clearly $b = 2 - a$.

We will show in the remaining cases that $b < a$. It is enough to check for $a = 1$.

If $A \cap E \neq \emptyset$ then since $A_1 \subset T$ contains at least two singularities, adjunction gives

$$0 > (K_T + A + bB) \cdot A \geq -1 + b.$$

So we can assume $x \notin A_1$. Then $T$ is not a net. $K_{S_2} + A + bB$ is negative, so again $b < 1$ by adjunction. □

For the inequalities and equalities for coefficients below, as well as the analysis in the exceptional case of the claim, we take $a = 1$.

**Claim.** *(8.4.7.6-11) hold*

*Proof.* $K_T + e_1 E + aA_1$ is level by (8.3.5.4). Since $b \leq a$, we can check flushness of $K_T + \Gamma'$ locally at each point of $E$, where it follows by (8.3.5.2). (8.4.7.6-11) now follow exactly as in the proof of (8.4.3). □

Note that $e_1 \geq 1/2$, and $e_1 \geq 2/3$ unless $A_1$ contains 4 singularities of index 2.

Consider the next hunt step. Let $j = -\tilde{E}^2$ (as in the usual hunt).

As $2a + b > 2$, we cannot have a tacnode, or a triple point, by (8.3.7).

Suppose $x \notin A$. If we go to a log del Pezzo, then $B$ is lt and smooth, since $A + B$ cannot have a triple point. Hence we must go to a fence. But as $A \subset S_2$ contains at least two singular points, $K_{S_2} + A + B$ is nef by adjunction, which contradicts (13.3). If $T$ is a net, then $A$ and $E$ are sections, for otherwise, since $2a + b > 2$, one is a section, and the other is a double section, which contradicts (11.5.11.2). There are at least 3 multiple fibres, so $x$ is a non-chain point. Now (1) follows from (23.4.1.1).

Suppose $x \in A$. We consider the various possibilities, (8.4.7.6-11).

If we have a Banana, then from the classification (13.2), (2-3) follow.

Now suppose $S_2$ is a fence. Since $A_1$ contains at least two singular points, by (13.3), $B$ is singular, $x \in A_1$ and $A_1$ is a $-1$-curve. Suppose $A_2$ does not lie in the Du Val locus. By considering the possible indices for singularities along $A_1$, and using the fact that $x$ has maximal index, it follows that $A_2$ contains either (3) or (4). One checks in all cases that $A_2^2 \leq 0$, a contradiction. Thus $B \subset S_2^0$, $g(B) = 1$, and $S_2$ is Du Val. Now the possibilities are classified in (13.4) and (13.5).

If $B$ has a cusp, then by (13.4) the singularities are $A_1 + A_2$. Consider $h : \tilde{T} \longrightarrow S_1$. Since $\tilde{E}_2^2 \leq -2$, the first five blow ups must be along $B$. This gives configuration $(v; n)$ of (11.2.1), and $x = (2)$. But then $r = 2$, and $A_2$ contains three $A_1$ points, contradicting (13.3). By (11.2.1) the configuration on $T$ is $(v; n^2)$, $x = (3)$ and $r = 3$. But then there is a point outside $A$ of coefficient 2/3, a contradiction, since by our conventions the hunt would choose that point.

Suppose $B$ has a node. Then $E$ contains an $A_r$ point, and if $E$ is a $-j$-curve, $-j+r+4 = K_{S_2}^2$. (4) follows by comparing the possible indices, with the possibilities of (13.5).

Suppose $x \in A$, and $T$ is a net. Then $A$ must be a fibre (otherwise by (11.5.11.2), as above, both $A$ and $E$ are sections, and $A$ contains at least two singular points, while $E$ contains at most one). By (11.5.5), $A_1$ contains exactly three singular points, so $e_1 \geq 2/3$.

Suppose $E$ has degree at least three. Then $E$ has degree exactly three, $A$ is a fibre of multiplicity three, $e_1 = 2/3$ and $r = 3$. $A \subset T$ is the fibre (11.5.9.5), with $k = 2$. In particular

$A \subset T$ contains singularities $A_2$ and (3). It follows from our convention for $x$, that $x = (3)$, and $E \subset T^0$. If we extract the $-3$-curve adjacent to $A$ and contract $A$, $E^2$ becomes zero, thus we obtain $\mathbb{F}_0$. This contradicts (11.5.10), since $E$ remains smooth and has degree three.

It follows $E$ is a double section and the singularities of $T$ along $A$ are each $A_1$. Consider $h : \tilde{T} \longrightarrow W$, for $W$ a smooth relative minimal model (see (11.5.4)). Note $K_T + E + \lambda F$ is trivial for some $\lambda > 0$. Thus $(K_W + E) \cdot E < 0$ and $E$ remains a smooth double section. By (11.5.10), $E^2 = 4$ on $W$, and $W = \mathbb{F}_1$ or $\mathbb{F}_0$. Let $p$, $q \in W$ be the ramification points of $E \longrightarrow \mathbb{P}^1$, with $p$ the point on the fibre corresponding to $A$. Clearly over $p$, $h$ is given by blowing up twice along $E$. Thus (since $\tilde{E}^2 \leq -2$ on $T$) $q$ must correspond to a singular fibre, $G$, of $T$. There can be no other singular fibres. The first $j+2$ blow ups of $h$ over $q$ must be along $E$. At this point $G$ is in the Du Val locus, and has a $D_j$ singularity. The next blow up (if it exists) is the non-interior blow up away from $E$ (see (11.5.4)), and subsequent blow ups are non-interior, determined by $x$. $S_1$ has a non-chain singularity $z$, with centre $-2$, and two branches (2). The last branch is determined by $x$. For example if $x = (\underline{2}, 3)$, then the last branch is $(3', 2)$, with the marked curve meeting the central curve. If $Z \longrightarrow S_1$ extracts the central divisor at $z$, then $Z$ is a net, $G$ is a fibre, $A$ and the central divisor are sections, and we have case (1).

The remaining possibility is that $A = \Sigma$. Thus $A$ is a $-1$-curve, and we have (8.4.7.7). $A_1 \subset T$ cannot contain two $A_1$ points (or it would not contract), so $b > 2/3$. It follows as in the proof of (18.7) that $B$ acquires an ordinary cusp, and the singularities along $\Sigma$ are (2) and (3).

If $B \subset S_2^0$, then $S_2$ is Du Val and $S_2^0$ is simply connected. $x = (j)$, $j \geq 3$. Then $S_2^2 = 6 - j \leq 3$. It follows from the simply connected list that $S_2 = S(E_{j+3})$. Let $Q$ be an end, opposite the central divisor, in an $A_{j-1}$ chain of the $E_{j+3}$ point. By (3.8), there is a unique $-1$-curve, $G$, meeting $Q$ (and no other curves over the singular point). Since $B \in |-K_{S_2}|$, $G$ meets $B$ normally. If $Z \longrightarrow S_1$ extracts the central divisor at $z$, then $G$ is a fibre, $A$ and the central divisor are sections, and we have case (1). In this case $G$ is the fibre (11.5.9) with $k = j - 1$.

Otherwise $B$ contains exactly one singular point $y$. Note by adjunction, $b = 5/6$. We consider the next hunt step with respect to $K_{S_2} + bB$.

We argue that $(S_3, B + C)$ is given by (13.4). If $x_2 \in B$, then $T = Y$ of (19.4), $M = \Sigma$, $B \subset S_3^0$ and the result is clear. So we can assume $x_2 \notin B$. Since $2b + c > 2$ we cannot have a net, or, by (8.3.7), a tacnode. $b + c > 1$, so we cannot have a triple point. Thus $S_3$ is a fence, and the result follows from (13.4).

Note $\tilde{B}_2^2 = 6 - j \leq 4$, while $B_3^2 = K_{S_3}^2 = 6$. Thus the first $j$ blow ups of $h : \tilde{T}_3 \longrightarrow S_3$ are along $B$. In particular $x_2 \notin B$, and $x_2$ is a non-chain singularity with center $-2$ and two branches (2) and (2, 2). Let $w = E_3 \cap \Sigma_3$. Note (see (11.1)) that $w$ is uniquely determined by $j$, and the marked point $y \in B$, and thus by $x \in A$. For example if $x = A_t$, then $w = (t+1)$. $x$ has index at most 6, and there are at least 2-curves over $x$ (since $B$ contains a singular point), and if $x$ is Du Val it has index at least 4 (or the hunt chooses the (3) point). Thus the possibilities for $x$ are $(\underline{3}, 2), (\underline{2}, 3)$ and $A_t$, with $3 \leq t \leq 5$. One checks for each possibility that $e(x_2) \geq \frac{r-1}{r}$, so with our convention, the hunt would choose $x_2$ instead of $x$, a contradiction. □

**23.5.1.1 Lemma.** *Let $(S_1, A_1)$ be as in (23.5). Suppose there is a non-chain singularity $z \notin A_1$ such that if $f : Z \longrightarrow S_1$ extracts the central curve, $E$, then $Z$ is a net and $E$ and $A_1$ are sections. Then:*

(1) *$(S_1, A_1)$ is uniquely determined by $z$.*
(2) *$K_{S_1} + A_1$ is negative.*
(3) *There are exactly three multiple fibres, each log terminal.*
(4) *Every non-chain singularity $z$ occurs for some pair $(S_1, A_1)$.*

*Proof.* Suppose $f : Z \longrightarrow S_1$ is as in the statement. Clearly there are three multiple fibres. Let $G$ be a multiple fibre. By (11.5.5), $G$ contains exactly two singular points, $G \cap E$ and $G \cap A_1$. The reduced components of $\mathcal{F}$ (in the language of (11.5.1) and (11.5.5)), each lie over one of the singular points. It follows there are no interior blow ups so $G$ is log terminal.

Let $h : \tilde{Z} \longrightarrow W$ be a relative minimal model which is an isomorphism in a neighbourhood of $\tilde{E} \subset \tilde{Z}$. By (11.5.7), $h$ is uniquely determined by $z$. If $E$ is a $-j$-curve, then $W = \mathbb{F}_j$, and $E \subset W$ generates one of the extremal rays, and $A_1 \subset E$ is an $j$-curve, disjoint from $E$. The first blow up of $h$ (for each multiple fibre) is along $A$, and then there is a sequence of interior blow ups uniquely determined by the marked singularity of $E$ at $E \cap G$. The singularities $A \cap G$ and $E \cap G$ have the same index, thus $K_{S_1} + A_1$ is negative (since $E$ is contractible). Any two $j$-curve in $\mathbb{F}_j$ disjoint from $E$ are permuted by an automorphism which is trivial on $E$. Thus $(S_1, A_1)$ is uniquely determined by $z$. □

*23.6 Remark.* In the case of (23.5.1) where $A$ contains two $A_1$ singularities, let $B \subset S_1$ be the image of the fibre which passes through the third singularity along $A$. Then $K_{S_1} + A + B$ is numerically trivial, log canonical, and $2(K_{S_1} + A + B)$ is linearly equivalent to zero. Thus $B$ is a two complement for $K_{S_1} + A$, see (22.3.2).

**23.7 The case (23.2.1).** Suppose $A_1$ contains at most two singularities. By (22.3.1) $K_{S_1} + A_1$ has a 1-complement, $X$. That is, there is a reduced curve $X$, containing $A$ as an irreducible component, with the following properties:

(*) $K_{S_1} + X$ is trivial, Cartier, log canonical, and $K_{S_1} + C$ is log terminal and anti-ample for every irreducible component $C$ of $X$.

We will forget about $A_1$, drop the subscript, and classify pairs $(S, X)$, satisfying (*).

*23.8 Remarks.* The conditions in (*), together with adjunction, imply the following:

(1) The only possible singularities of $S$ along $X$ are at the intersection points of two irre-

ducible components.

(2) $X$ is necessarily reducible, and has either two irreducible components, meeting in two points, or has three irreducible components forming a triangle.

(3) $S \setminus X$ is Du Val.

We first introduce two useful birational transformations. We start with a pair $(S, X)$ and obtain a new pair $(S', X')$. $S'$ is obtained by one blow up and one blow down, and $X'$ consists of the pushforward of the total transform of $X$.

We remind the reader of our sporadically employed convention of using the same notation to indicate a curve, and its strict transform under a birational transformation.

**23.9 Definition-Lemma.**

*(a) Suppose a component, $C$, of $X$ is a $-1$-curve. Let $T \longrightarrow S$ extract a divisor, $E$, adjacent to $C$. $C \subset T$ is contractible. Let $T \longrightarrow S'$ be the contraction of $C$. Let $q \in S'$ be the image of $C$.*

*(b) Suppose there a $-1$-curve $\Sigma$ with the following properties:*

*(1) $\Sigma \cap X$ is a single point, $p$, a singular point of $S$, and $\Sigma$ meets a unique curve of the minimal desingularisation over $p$.*

*(2) The strict transforms of $\Sigma$ and $X$ on the minimal desingularisation are disjoint.*

*(3) $\Sigma$ contains at most one other singular point of $S$, and $K_S + \Sigma$ is log terminal away from $p$.*

*Let $T \longrightarrow S$ extract the divisor, $E$, over $p$ which is adjacent to $\Sigma$. $\Sigma \subset T$ is contractible. Let $T \longrightarrow S'$ be the contraction of $\Sigma$. Let $q \in S'$ be the image of $\Sigma$.*

*In both cases the new pair $(S', X')$ also satisfies (*). In case (a), $X'$ has the same number of components as $X$, but in case (b) the number of components goes up by one.*

The inverse transformations $(S', X') \dashrightarrow (S, X)$ are easy to describe:

**23.10 Definition-Lemma.** *Let $(S', X')$ satisfy (*). Let $q \in X'$ be a singular point (resp. a smooth point) of $X'$, lying on an irreducible component $E$ of $X$ (there are two choices of $E$ when $q$ is a singular point, and only one choice when $q$ is a smooth point). Let $\tilde{S} \longrightarrow \tilde{S}'$ be a composition of blow ups, with the center of each blow up at the unique point of (the strict transform of) $E$, lying over $q$. Blow up at least enough times so that the self-intersection of $E \subset \tilde{S}$ is at most $-2$. Let $C \subset \tilde{S}$ (resp. $\Sigma \subset \tilde{S}$) be the unique $-1$-curve over $p$, that is the exceptional divisor of the last blow up. Let $\tilde{S} \longrightarrow S$ contract all the $K$ non-negative curves, one of which, by construction, is $E$. Let $p \in S$ be the image of $E \subset \tilde{S}$. $(S, X)$ satisfies (*), and $(S', X')$ is obtained from $(S, X)$ by making a transformation of type (a) (resp (b)) extracting $E$, and using $p \in C \subset S$ (resp. $p \in \Sigma \subset S$).*

*Proofs of (23.9-10).* These are easy, see for example (11.4). $\square$

We will refer to the transformations of (23.9-10) as being of type (a),(b),$(a)^{-1}$ and $(b)^{-1}$.

**23.11 Lemma.** *Let $\overline{S} \longrightarrow S$ be the minimal desingularisation in a neighbourhood of $X$ (that is resolving only the singularities along $X$). If $\overline{S}$ has a $K_{\overline{S}}$-negative birational contraction, then $(S, X)$ has a transformation of type (a) or (b).*

*Proof.* Note by (23.8.3), $\overline{S}$ is Du Val. Suppose $\overline{S}$ has a $K_{\overline{S}}$-negative birational contraction. If the associated $-1$-curve is a component, $C$ of $X$, then clearly we have a transformation of type (a). Otherwise call the $-1$-curve $\Sigma$. By (3.3), $K_{\overline{S}}+\Sigma$ is log terminal, and $\Sigma$ contains at most one singularity. Let $X' \subset \overline{S}$ be the total transform of $X$. $K_{\overline{S}} + X' = \pi^*(K_S + X)$, thus $K_{\overline{S}} + X'$ is numerically trivial. Thus $X' \cdot D = -K_{\tilde{S}} \cdot \tilde{\Sigma} = 1$. Thus $\Sigma$ meets a unique irreducible component of $X'$. This must be an exceptional component, or $\Sigma \subset S$ is contractible. □

**23.12 Proposition.** *Let $(S, X)$ satisfy (*). One of the following holds:*

*(1) The pair $(S, X)$ is toric (see (22.4)), or obtained from a toric pair by a single transformation of type $(b)^{-1}$.*
*(2) $S = \overline{\mathbb{F}}_n$, $X$ consists of a fibre union a section which passes through the singular point. (As a special case we get $\mathbb{P}^2$ and $X$ is a conic union a line.)*
*(3) $S = \overline{\mathbb{F}}_2$ and $X$ consists of two sections in the smooth locus.*
*(4) $(S, X)$ is obtained from the pair in (2) by a sequence of transformations of type $(a)^{-1}$. Moreover, exactly one component, $C$, of $X$ is a $-1$-curve and $C$ has a 1-complement $Y$ (see §22) such that $(S, Y)$ is toric.*

*23.13 Remarks.*

(1) In case (4), the proof gives a bit more information. For the transformations, the section is never contracted. $C$ is the other component.
(2) Note in particular that the surface $S$ is toric, except possibly in the second case of (1).

*Proof.* Note there can only be a finite sequence of transformations (of any type), since each improves the singularities. Note also, in any sequence of transformations, there can be at most one of type (b), since otherwise $X'$ would have at least four components, violating (23.8.1). Finally note that if $(S', X')$ has a transformation of type (b), then so does $(S, X) = (a)^{-1}(S', X')$ (an $(a)^{-1}$ transformation will not destroy a curve of type $\Sigma$ in (23.9)).

Now start with a pair $(S, X)$. Perform a transformation of type (b) if one exists. The only further possible transformations are of type (a).

Now assume we have $(S, X)$ and there is no transformation of type (a) or (b). Then by (23.7), either $S$ is smooth along $X$, or $\overline{S}$, the minimal desingularisation of $S$ in a neighbourhood of $X$, is a net, of Picard number two. Since $K_{\overline{S}} + X'$ is trivial, and $X' \subset \overline{S}^0$, it follows that $\overline{S}$ is smooth. Now if $X$ has two components it is easy to see we have (2) or (3), and if $X$ has three components, it is easy to check that $S = \overline{\mathbb{F}}_n$ and $X$ is two fibres union a section in the smooth locus (as a special case we get $\mathbb{P}^2$ and three lines), which are all toric.

Note $(a)^{-1}$ transformations preserve toric pairs. So if we end with three components, we began in the situation of (1).

So we can assume we end with (2) or (3), and we obtain the original pair by a sequence of $(a)^{-1}$ transformations.

In (3), after one transformation of type $(a)^{-1}$, there is a curve of type $\Sigma$ (the strict transform of the fibre through $q$), that is the new pair admits a transformation of type (b). We ruled out this possibility above.

So we can assume our original pair is obtained from $(a)^{-1}$ transformations applied to (2). We need only check the extra conditions in (4). Let $X_1$ be the section of self-intersection $n+2$, and $X_2$ the fibre.

Let $y \in \mathbb{F}_n$ be the intersection point of the strict transforms of $X_1$ and $X_2$. Let $x \in \mathbb{F}_n$ be the point where $X_1$ meets the negative section. Let $G$ be the fibre of $\overline{\mathbb{F}}_n$ through $x$. There is a unique section $\Sigma_n \subset \mathbb{F}_n$ which is disjoint from the negative section, and has $(n+1)^{\text{st}}$ order contact with $X_1$ at $y$. Scheme-theoretically, $X_1 \cap \Sigma_n = (n+1)y$. As remarked above, the pair $(X_2 + G + \Sigma_n, \overline{\mathbb{F}}_n)$ is toric.

We will argue that after any blow up $\tilde{X}_1^2$ (the self-intersection of the strict transform on the minimal desingularisation) remains positive. In particular, $X_1$ will never be chosen as $E$ (in the language of (23.10)). Thus each of the transformations is toric with respect to $(X_2 + G + \Sigma_n)$ (and its transforms). (4) then follows immediately.

Note that unless $q$ (in the language of (23.10)) corresponds to either $x$ or $y$, there is a neighbourhood of $\tilde{G} + \tilde{\Sigma}_n + \tilde{X}_1$ (on the minimal desingularisation) which is not effected. Further there is a neighbourhood of $\tilde{G}$ which is effected only if $q = x$, and any choice of $q$ has the same effect on numbers $\tilde{X}_1^2$, $\tilde{\Sigma}_n^2$, and $X_1 \cdot \Sigma_n$ (the order of contact at $y$).

Now if we choose $q = x$, then after the transformation $G$ is a $-1$-curve of type $\Sigma$, a possibility we have ruled out.

If after a blow up (in the construction of $\tilde{S} \longrightarrow \tilde{S}'$ in (23.10)) $\tilde{X}_1$ becomes non-positive, then $\tilde{\Sigma}_n$ is separated from $\tilde{X}_1$, and on the new surface, $\Sigma_n$ is a $-1$-curve of type $\Sigma$, a contradiction as before. $\square$

## Appendix $L$: Log terminal surface singularities and adjunction

We note for the readers convenience (or edification) that a normal surface is log terminal iff it has quotient singularities, and that for $C$ an analytically irreducible germ of a curve at $p$, $p \in C \subset S$, if $S$ is given locally analytically about $p$ as the quotient $q : V \longrightarrow S$ with $V$ smooth, then $K_S + C$ is log terminal (resp. log canonical) at $p$ iff $q^*(C) \subset V$ is smooth (resp. has normal crossings). For a proof see (20.3) of [27].

Let $p \in C \subset S$ be a point of $S$ on the (non-empty) reduced curve $C$. We give the criteria for log terminal and log canonical at $p$ in this case. For proofs see chapter 3 of [27].

Suppose first $p$ is a singular point of $S$:

$K_S + C$ is lt iff $p$ is a cyclic singularity, and the resolution graph is a chain, with $\tilde{C}$ meeting one end of the chain, and normally. In particular $C$ is smooth at $p$. We will refer to the unique exceptional divisor which $\tilde{C}$ meets as the divisor **adjacent to** $C$ over $p$.

If $C$ is singular at $p$, then $K_S + C$ is log canonical (lc) at $p$ iff $C$ has two analytic branches at $p$, each of which if smooth, $p$ is a chain singularity, the two branches of $\tilde{C}$ are disjoint, and meet opposite ends of the chain, and normally.

If $C$ is smooth at $p$ then $K_S + C$ is lc but not lt, iff $p$ is a non-chain singularity, with two branches (2), (2), and $\tilde{C}$ meets normally the end of third branch, at the opposite end from the central divisor.

Note in particular that the criteria above are local analytic. When $p$ is smooth this is not the case:

$K + C$ is lt at $p$ iff either $C$ is smooth at $p$, or $C$ has two irreducible components (in a Zariski neighbourhood), each smooth, meeting normally at $p$.

$K + C$ is lc at $p$ but not lt, iff $C$ is irreducible (in a Zariski neighbourhood of $p$), and has a simple node at $p$.

Thus for example if $C$ is the union of three general lines in $\mathbb{P}^2$, $K + C$ is lt, but if $C$ is a nodal cubic, it is lc but not lt.

Because of this peculiarity of the definition, we will sometimes use the variant purely log terminal (plt). For the general definition see chapter 2 of [27]. For our purposes it is enough that $K + C$ is plt at a singular point $p$ iff it is lt at $p$, and plt at a smooth point $p$ iff $C$ is smooth at $p$. In short, $K + C$ is plt iff it is lt and $C$ is smooth.

We often use the following standard result implicitly in this paper:

**L.1 Lemma.** *Let $(S, x)$, $(T, y)$ be germs of log terminal surface singularities, and let $\Gamma', \Gamma$ be the dual graphs of their minimal resolutions. Assume $\Gamma$ is obtained from $\Gamma'$ either by increasing the weight of one of the vertices, or by adding a single vertex and edge. The following hold:*

(1) *The index of $y$ is strictly bigger than the index of $x$.*

(2) *The coefficient of each exceptional divisor of the minimal resolution of $y$ is strictly bigger than the coefficient of the corresponding divisor over $x$, unless $y$ is a Du Val singularity (in which case all the coefficients are zero).*

*Proof.* The cyclic case of (1) follows from (3.1.8) of [27]. The non-cyclic case (and indeed the cyclic case as well) can be checked from the explicit lists in [7].

For (2) we will consider the case when the weight goes up. The other case is handled analogously, or alternatively, follows from (3.1.3) of [27].

Of course we may assume the weight goes up by exactly one. Since the singularities are determined by the resolution graph, we can assume $(T, y)$ is obtained from $(S, x)$ by a smooth blow up $h : \tilde{T} \longrightarrow \tilde{S}$ at a point lying on a unique exceptional divisor, the divisor corresponding to the vertex whose weight increases. Let $\Sigma$ be the $-1$-curve extracted by $h$. Note $e = e(\Sigma, K_S) < 0$ since $y$ is log terminal. Let $E$ be an exceptional divisor (of the minimal desingularisation) over $x$. Let $f : T' \longrightarrow T$, $g : S' \longrightarrow S$ extract $E$. We have an induced map $h' : T' \longrightarrow S'$.

$$(g \circ h')^*(K_S) = h'^*(K_{S'} + e(E, K_S)E) = K_{T'} + e(E, K_S)E + (e-1)\Sigma.$$

Thus, computing intersections on $T'$,

$$-e(E, K_S)E^2 = (K_{T'} + (e-1)\Sigma) \cdot E < K_T' \cdot E = (f^*(K_T) - e(E, K_T)E) \cdot E = -e(E, K_T)E^2. \quad \square$$

## L.2 Adjunction and Inversion of Adjunction.

In chapter 3 of [34], Shokurov introduces the following:

**Definition-Lemma (The Different).** Let $S$ be a normal variety and $C \subset S$ a reduced effective divisor. Let $\overline{C}$ be the normalisation, and $i : \overline{C} \longrightarrow S$ the induced map. Let $B$ be any effective $\mathbb{Q}$-divisor on $S$. Assume $K_S + C + B$ is $\mathbb{Q}$-Cartier:

There is a canonically defined effective $\mathbb{Q}$-Weil divisor $\mathrm{Diff}_C(B)$ on $C$ such that

$$i^*(K_S + C + B) = K_C + \mathrm{Diff}_C(B)$$

(The assumption that $K_S + C + B$ is $\mathbb{Q}$-Cartier is not necessary, but without it the definition of $i^*$ is more technical. The above is sufficient for our needs.)

When $(S, C)$ is log terminal the different has a simple interpretation via Kollár's Bug-Eyed cover. See (4.15).

Shokurov obtained the following quite surprising result:

**L.2.1 Theorem (Inversion of Adjunction).** *Assume* $\dim(S) \leq 3$.

*$K_S + C + B$ is log canonical (resp. plt and $B$ is a pure boundary) in a neighbourhood of $C$ iff $K_C + \mathrm{Diff}_C(B)$ is log canonical (resp. klt).*

For the proof see (3.4) of [34],

By an elegant extension of an idea of Shokurov, Kollár proved the plt case of (L.2.1) in all dimensions, see (17.6) of [27].

**L.2.2 Lemma.** *Let $C \subset S$ be a reduced irreducible curve on a $\mathbb{Q}$-factorial projective surface. If $(K_S + C) \cdot C < 0$, then $C = \mathbb{P}^1$.*

*Proof.* By the subadjunction formula, (5.1.9) of [21], for sufficiently divisible $m > 0$, $\omega_C^{[m]}$ is a subsheaf of the line bundle $\mathcal{O}_C(m(K_S + C))$. As the latter has negative degree, it follows that $H^1(\mathcal{O}_C) = H^0(\omega_C) = 0$. □

*Remark.* We expect there is also a local version of (L.2.2), namely that $\mathrm{Diff}_C(0)$ is at least as big as the conductor, which is either empty, or has degree at least two.

The next result is our principal application of (L.2.1):

**L.2.3 Corollary.** *Let $S$ be a $\mathbb{Q}$-factorial projective surface, and $C, D \subset S$ two curves, with $C$ integral. If $(K_S + C + D) \cdot C \leq 0$, and $C \cap D$ contains at least two points, then $K_S + C + D$ is log canonical, $C \cap D$ is exactly two points, and $(K_S + C + D) \cdot C = 0$.*

*Proof.* $C = \mathbb{P}^1$ by (L.2.2). By (L.2.1), since $K + C + tD$ is not lc at $D \cap C$, for any $t > 1$, the coefficient of $Q$ in $\mathrm{Diff}_C(D)$ is at least one, for any $Q \in D \cap C$. It follows that $D \cap C$ is exactly two points $\{Q_1, Q_2\}$ and $\mathrm{Diff}_C(D) = Q_1 + Q_2$. In particular $K_C + \mathrm{Diff}_C(D)$ is lc, and thus $K_S + C + D$ is lc in a neighbourhood of $C$ by (L.2.1). □

**Warning:** In this paper we frequently use (L.2.1-3) without reference, or under the catch all *by adjunction.*

## Appendix N: Normalisation of an algebraic space

Here we give a construction of the normalisation of an algebraic space, as we have not been able to find it in the literature.

Note that the property of being normal is preserved by étale covers, and so makes sense for algebraic spaces.

Recall for a reduced scheme $X$ there is a unique finite morphism $\tilde{X} \longrightarrow X$ (the normalisation) which is universal for dominant maps from normal schemes.

**N.1 Lemma.** *Let $f : Y \longrightarrow X$ be a dominant étale map of schemes, with $X$ reduced. Then there is a natural fibre diagram*

$$\begin{array}{ccc} \tilde{Y} & \longrightarrow & Y \\ \downarrow & & \downarrow \\ \tilde{X} & \longrightarrow & X. \end{array}$$

*Proof.* Immediate from the universal property of the normalisation. □

Now let $j : R \longrightarrow Y \times Y$ be an étale equivalence relation. By the universal property there is an induced map $\tilde{j} : \tilde{R} \longrightarrow \tilde{Y} \times \tilde{Y}$.

**N.2 Lemma.** *(assumptions as above)* $\tilde{j}$ *is an étale equivalence relation.*

*Proof.* By the definition of an equivalence relation, we need to show that $j$ is a monomorphism, and $\operatorname{Hom}(T, \tilde{R}) \subset \operatorname{Hom}(T, \tilde{Y}) \times \operatorname{Hom}(T, \tilde{Y})$ is an equivalence relation for all $T$.

It follows from (N.1) that $\tilde{j}$ is a monomorphism, and the projections $\tilde{R} \longrightarrow \tilde{Y}$ are étale. Also, by the universal property, for any normal scheme $T$, $\operatorname{Hom}_d(T, \tilde{R}) = \operatorname{Hom}_d(T, R)$ and $\operatorname{Hom}_d(T, \tilde{Y} \times \tilde{Y}) = \operatorname{Hom}_d(T, Y \times Y)$, where $\operatorname{Hom}_d$ indicates the subset of dominant maps. It follows that $\operatorname{Hom}_d(T, \tilde{R}) \subset \operatorname{Hom}_d(T, \tilde{Y}) \times \operatorname{Hom}_d(T, \tilde{Y})$ is an equivalence relation for any normal $T$.

For the transitivity axiom. Let $T = \tilde{R} \underset{(p_2, p_1)}{\times} \tilde{R}$, where $p_1, p_2 : \tilde{R} \longrightarrow \tilde{Y}$ are the two projections. It is enough to show that $(p_1 \circ pr_1, p_2 \circ pr_2) \in \operatorname{Hom}(T, \tilde{R})$. But since $T$ is normal, and $p_i$ are dominant, this follows from the remarks above. The other axioms are analogously obtained. □

**N.3 Definition-Lemma: Normalisation.** *Let $W$ be a reduced algebraic space. Then there is a unique finite dominant map $\tilde{W} \longrightarrow W$ from a normal algebraic space, universal for dominant maps from normal algebraic spaces.*

*Proof.* Let $f : Y \longrightarrow W$ be an étale cover by an affine scheme, let $R = Y \underset{W}{\times} Y$, so that $W = Y/R$. Let $\tilde{W} = \tilde{Y}/\tilde{R}$, which is an algebraic space by (N.2), and clearly normal. By construction there is an induced map $\tilde{W} \longrightarrow W$. (N.1) and the construction implies $\tilde{Y} = \tilde{W} \underset{W}{\times} Y$. Now the universal property of $\tilde{Y} \longrightarrow Y$ implies the universality of $\tilde{W} \longrightarrow W$. □

*N.4 Remark.* Note that Lemma (N.1) holds for algebraic spaces, just copy the proof of (N.1) and use (N.3).

## Index

## References


[1] V. A. Alekseev and V. V. Nikulin. Classification of del Pezzo surfaces with log-terminal singularties of index $\leq 2$ and involutions on $K3$ surfaces. *Soviet Math. Dokl.*, 39:507–511, 1989.

[2] V. Alexeev. Boundedness and $K^2$ for log surfaces. *International J. Math.*, 5:779–810, 1994.

[3] M. Artin. Algebraic approximation of structures over complete local rings. *Pb. Math. I.H.E.S*, 36:25–58, 1969.

[4] W. Barth, C. Peters, and A. Van de Ven. *Compact Complex Surfaces.* Springer Verlag, 1984.

[5] V. Batyrev. The cone of effective divisors of threefolds. In *Proc. Intl. Conf. on Algebra*, pages 337–352, 1989.

[6] A. Borisov. Boundedness theorem for Fano log-threefolds. *J. Algebraic Geom.*, 5:119–133, 1996.

[7] E. Brieskorn. Rationale Singulariten komplexer Flachen. *Invent. Math.*, 4:336–358, 1968.

[8] A. Fujiki, R. Kobayishi, and S. Luo. On the fundamental group of certain open normal surfaces. *Saitama Math. J.*, 11:15–20, 1993.

[9] W. Fulton. *Intersection Theory*. Spinger-Verlag, 1984.

[10] W. Fulton. *Introduction to toric varieties*. Princeton University Press, 1993.

[11] M. Furushima. Singular del Pezzo surfaces and analytic compacitifications of 3-dimensional complex affine space $\mathbb{C}^3$. *Nagoya Math. J.*, 104:1–28, 1986.

[12] Mark L. Green. *Infinitesimal methods in Hodge theory*. Springer-Verlag, 1994. Lecture Notes in Math. vol. 1594.

[13] A. Grothendieck. Eléménts de Géométrie Algébrique. *Publ. Math. IHES*, 28, 1966.

[14] R. V Gurjar and D. Zhang. $\pi_1$ of smooth points of a log del Pezzo surface is finite:II. MPI preprint, 1993.

[15] R. V Gurjar and D. Zhang. $\pi_1$ of smooth points of a log del Pezzo surface is finite:I. *J. Math. Sci. Univ. Tokyo*, 1:137–180, 1994.

[16] H.Clemens, J.Kollár, and S.Mori. Higher dimensional complex geometry. *Astérisque, Soc. Math. France,*, 166, 1988.

[17] F. Hidaka and K. Watanabe. Normal Gorenstein surfaces with ample anticanonical divisor. *Tokyo J. Math.*, 4:283–360, 1981.

[18] Y. Miyaoka J. Kollár and S. Mori. Rational curves on Fano varieties. *J. Diff. Geom.*, 3:765–779, 1992.

[19] Y. Miyaoka J. Kollár and S. Mori. Rationally connected varieties. *J. Algebraic Geometry*, 1:429–448, 1992.

[20] Y. Kawamata. Boundedness of $\mathbb{Q}$-Fano threefolds. *Contemp. Math., Part 3, Amer. Math. Soc.*, 131, 1992.

[21] Y. Kawamata, K. Matsuda, and K. Matsuki. Introduction to the minimal model program. In T. Oda, editor, *Algebraic Geometry, Sendai*, pages 283–360. Kinokuniya-North-Holland, 1987. Adv. Stud. Pure Math, vol 10.

[22] S. Keel, K. Matsuki, and J. McKernan. Log abundance theorem for threefolds. *Duke Journal of Mathematics*, 1994.

[23] J. Kollár. Cone theorems and bug-eyed covers. *J. Algebraic Geometry*, 1:293–323, 1992.

[24] J. Kollár. Shafarevich maps and automorphic forms. Lectures at Rice University, 1993.

[25] J. Kollár. Log surfaces of general type; some conjectures. *Contemp. Math.*, 162:261–275, 1994.

[26] J. Kollár. *Rational Curves on Algebraic Varieties.* Springer Verlag, 1996.

[27] J. Kollár et al. Flips and abundance for algebraic threefolds. *Astérisque*, 211, 1992.

[28] H. Matsummura. *Commutative Algebra.* Benjamin-Cummings, 1980.

[29] V. B. Mehta and A. Ramanathan. Semi–stable sheaves on projective varieties and their restriction to curves. *Math. Ann.*, 258:213–224, 1982.

[30] M. Miyanishi and S. Tsunoda. The structure of open algebraic surfaces, II. In *Classification of algebraic and analytic manifolds (Katata. 1982)*, pages 499–544. Birkhauser, 1983. Progr. Math. vol 39.

[31] M. Miyanishi and S. Tsunoda. Logarithmic del Pezzo surfaces of rank one with contractible boundaries at infinity. *Japan J. Math.*, 10:271–319, 1984.

[32] M. Miyanishi and S. Tsunoda. Non complete algebraic surface with logarithmic Kodaira dimension $-\infty$ and with non-connected boundaries at infinity. *Japan J. Math.*, 10:195–242, 1984.

[33] M. Miyanishi and D. Q. Zhang. Gorenstein log del Pezzo surfaces of rank one. *Journal of Algebra*, 118:63–84, 1988.

[34] V. V. Shokurov. 3–fold log flips. *Russian Acad. Sci. Izv. Math.*, 40(1), 1993.

[35] D. Zhang. Algebraic surfaces with nef and big anti-canonical divisor. *Math. Proc. Cambridge Philos. Soc.*, 117:161–163, 1995.

[36] Qe-Qi Zhang. Logarithmic del Pezzo surfaces of rank one with contractible boundaries. *Osaka J. Math.*, 25:461–497, 1988.

## Editorial Information

To be published in the *Memoirs*, a paper must be correct, new, nontrivial, and significant. Further, it must be well written and of interest to a substantial number of mathematicians. Piecemeal results, such as an inconclusive step toward an unproved major theorem or a minor variation on a known result, are in general not acceptable for publication. *Transactions* Editors shall solicit and encourage publication of worthy papers. Papers appearing in *Memoirs* are generally longer than those appearing in *Transactions* with which it shares an editorial committee.

As of March 31, 1999, the backlog for this journal was approximately 4 volumes. This estimate is the result of dividing the number of manuscripts for this journal in the Providence office that have not yet gone to the printer on the above date by the average number of monographs per volume over the previous twelve months, reduced by the number of issues published in four months (the time necessary for preparing an issue for the printer). (There are 6 volumes per year, each containing at least 4 numbers.)

A Copyright Transfer Agreement is required before a paper will be published in this journal. By submitting a paper to this journal, authors certify that the manuscript has not been submitted to nor is it under consideration for publication by another journal, conference proceedings, or similar publication.

## Information for Authors and Editors

*Memoirs* are printed by photo-offset from camera copy fully prepared by the author. This means that the finished book will look exactly like the copy submitted.

The paper must contain a *descriptive title* and an *abstract* that summarizes the article in language suitable for workers in the general field (algebra, analysis, etc.). The *descriptive title* should be short, but informative; useless or vague phrases such as "some remarks about" or "concerning" should be avoided. The *abstract* should be at least one complete sentence, and at most 300 words. Included with the footnotes to the paper, there should be the 1991 *Mathematics Subject Classification* representing the primary and secondary subjects of the article. This may be followed by a list of *key words and phrases* describing the subject matter of the article and taken from it. A list of the numbers may be found in the annual index of *Mathematical Reviews*, published with the December issue starting in 1990, as well as from the electronic service e-MATH [**telnet e-MATH.ams.org** (or **telnet 130.44.1.100**). Login and password are **e-math**]. For journal abbreviations used in bibliographies, see the list of serials in the latest *Mathematical Reviews* annual index. When the manuscript is submitted, authors should supply the editor with electronic addresses if available. These will be printed after the postal address at the end of each article.

**Electronically prepared papers.** The AMS encourages submission of electronically prepared papers in $\mathcal{A}_{\mathcal{M}}\mathcal{S}$-TeX or $\mathcal{A}_{\mathcal{M}}\mathcal{S}$-LaTeX. The Society has prepared author packages for each AMS publication. Author packages include instructions for preparing electronic papers, the *AMS Author Handbook*, samples, and a style file that generates the particular design specifications of that publication series for both $\mathcal{A}_{\mathcal{M}}\mathcal{S}$-TeX and $\mathcal{A}_{\mathcal{M}}\mathcal{S}$-LaTeX.

Authors with FTP access may retrieve an author package from the Society's Internet node `e-MATH.ams.org` (130.44.1.100). For those without FTP access, the author package can be obtained free of charge by sending e-mail to `pub@ams.org` (Internet) or from the Publication Division, American Mathematical Society, P.O. Box 6248, Providence, RI 02940-6248. When requesting an author package, please specify $\mathcal{A}\mathcal{M}\mathcal{S}$-TeX or $\mathcal{A}\mathcal{M}\mathcal{S}$-LaTeX, Macintosh or IBM (3.5) format, and the publication in which your paper will appear. Please be sure to include your complete mailing address.

**Submission of electronic files.** At the time of submission, the source file(s) should be sent to the Providence office (this includes any TeX source file, any graphics files, and the DVI or PostScript file).

Before sending the source file, be sure you have proofread your paper carefully. The files you send must be the EXACT files used to generate the proof copy that was accepted for publication. For all publications, authors are required to send a printed copy of their paper, which exactly matches the copy approved for publication, along with any graphics that will appear in the paper.

TeX files may be submitted by email, FTP, or on diskette. The DVI file(s) and PostScript files should be submitted only by FTP or on diskette unless they are encoded properly to submit through e-mail. (DVI files are binary and PostScript files tend to be very large.)

Files sent by electronic mail should be addressed to the Internet address `pub-submit@ams.org`. The subject line of the message should include the publication code to identify it as a Memoir. TeX source files, DVI files, and PostScript files can be transferred over the Internet by FTP to the Internet node `e-math.ams.org` (130.44.1.100).

**Electronic graphics.** Figures may be submitted to the AMS in an electronic format. The AMS recommends that graphics created electronically be saved in Encapsulated PostScript (EPS) format. This includes graphics originated via a graphics application as well as scanned photographs or other computer-generated images.

If the graphics package used does not support EPS output, the graphics file should be saved in one of the standard graphics formats—such as TIFF, PICT, GIF, etc.—rather than in an application-dependent format. Graphics files submitted in an application-dependent format are not likely to be used. No matter what method was used to produce the graphic, it is necessary to provide a paper copy to the AMS.

Authors using graphics packages for the creation of electronic art should also avoid the use of any lines thinner than 0.5 points in width. Many graphics packages allow the user to specify a "hairline" for a very thin line. Hairlines often look acceptable when proofed on a typical laser printer. However, when produced on a high-resolution laser imagesetter, hairlines become nearly invisible and will be lost entirely in the final printing process.

Screens should be set to values between 15% and 85%. Screens which fall outside of this range are too light or too dark to print correctly.

Any inquiries concerning a paper that has been accepted for publication should be sent directly to the Editorial Department, American Mathematical Society, P. O. Box 6248, Providence, RI 02940-6248.

## Editors

## Selected Titles in This Series

*(Continued from the front of this publication)*

640 **Xingde Dai and David R. Larson,** Wandering vectors for unitary systems and orthogonal wavelets, 1998

639 **Joan C. Artés, Robert E. Kooij, and Jaume Llibre,** Structurally stable quadratic vector fields, 1998

638 **Gunnar Fløystad,** Higher initial ideals of homogeneous ideals, 1998

637 **Thomáš Gedeon,** Cyclic feedback systems, 1998

636 **Ching-Chau Yu,** Nonlinear eigenvalues and analytic-hypoellipticity, 1998

635 **Magdy Assem,** On stability and endoscopic transfer of unipotent orbital integrals on $p$-adic symplectic groups, 1998

634 **Darrin D. Frey,** Conjugacy of $\mathrm{Alt}_5$ and $\mathrm{SL}(2,5)$ subgroups of $\mathrm{E}_8(\mathbb{C})$, 1998

633 **Dikran Dikranjan and Dmitri Shakhmatov,** Algebraic structure of pseudocompact groups, 1998

632 **Shouchuan Hu and Nikolaos S. Papageorgiou,** Time-dependent subdifferential evolution inclusions and optimal control, 1998

631 **Ronnie Lee, Steven H. Weintraub, and J. William Hoffman,** The Siegel modular variety of degree two and level four/Cohomology of the Siegel modular group of degree two and level four, 1998

630 **Florin Rădulescu,** The $\Gamma$-equivariant form of the Berezin quantization of the upper half plane, 1998

629 **Richard B. Sowers,** Short-time geometry of random heat kernels, 1998

628 **Christopher K. McCord, Kenneth R. Meyer, and Quidong Wang,** The integral manifolds of the three body problem, 1998

627 **Roland Speicher,** Combinatorial theory of the free product with amalgamation and operator-valued free probability theory, 1998

626 **Mikhail Borovoi,** Abelian Galois cohomology of reductive groups, 1998

625 **George Xian-Zhi Yuan,** The study of minimax inequalities and applications to economies and variational inequalities, 1998

624 **P. Deift and K. T-R McLaughlin,** A continuum limit of the Toda lattice, 1998

623 **S. A. Adeleke and Peter M. Neumann,** Relations related to betweenness: Their structure and automorphisms, 1998

622 **Luigi Fontana, Steven G. Krantz, and Marco M. Peloso,** Hodge theory in the Sobolev topology for the de Rham complex, 1998

621 **Gregory L. Cherlin,** The classification of countable homogeneous directed graphs and countable homogeneous $n$-tournaments, 1998

620 **Victor Guba and Mark Sapir,** Diagram groups, 1997

619 **Kazuyoshi Kiyohara,** Two classes of Riemannian manifolds whose geodesic flows are integrable, 1997

618 **Karl H. Hofmann and Wolfgang A. F. Ruppert,** Lie groups and subsemigroups with surjective exponential function, 1997

617 **Robin Hartshorne,** Families of curves in $\mathbb{P}^3$ and Zeuthen's problem, 1997

616 **Serguei G. Bobkov and Christian Houdré,** Some connections between isoperimetric and Sobolev-type inequalities, 1997

615 **Michael A. Dritschel and Hugo J. Woerdeman,** Model theory and linear extreme points in the numerical radius unit ball, 1997

614 **Richard Warren,** The structure of $k$-$CS$-transitive cycle-free partial orders, 1997

613 **D. L. Flannery,** The finite irreducible linear 2-groups of degree 4, 1997

612 **Joan Porti,** Torsion de Reidemeister pour les variétés hyperboliques, 1997

611 **D. Ginzburg, I. Piatetski-Shapiro, and S. Rallis,** $L$ functions for the orthogonal group, 1997

(See the AMS catalog for earlier titles)